Telluride Ores of
Boulder County, Colorado

The Memoir Series
of
The Geological Society of America
is made possible
through the generous contributions of
Richard Alexander Fullerton Penrose, Jr.,
and is partially supported by
the National Science Foundation.

The Geological Society of America, Inc.
Memoir 109

Telluride Ores of
Boulder County, Colorado

WILLIAM C. KELLY
EDWIN N. GODDARD

Department of Geology and Mineralogy,
The University of Michigan,
Ann Arbor, Michigan

1969

PUBLISHED BY

THE GEOLOGICAL SOCIETY OF AMERICA, INC.
Colorado Building, P.O. Box 1719
Boulder, Colorado 80302

Printed in the United States of America

Acknowledgments

The study was carried out in the laboratories of the Department of Geology and Mineralogy at The University of Michigan, and was generously sponsored by grants from the Horace H. Rackham Faculty Research Fund. Equipment and facilities furnished by the National Science Foundation also made this project possible.

Many colleagues and friends in the department contributed to the progress of our work, and we are especially indebted to Professor F. Stewart Turneaure for his continuing interest and stimulating discussions; to Professor Louis I. Briggs and Dr. Frank Moser who assisted in computer analysis of assay data; to Professor E. William Heinrich who kept a watchful eye for literature pertaining to our studies; and to Professors Paul L. Cloke and Donald R. Peacor, who spent long hours in active discussion of the geochemistry and mineralogy of the telluride ores. The illustrations in this paper were skillfully drafted by Mr. Derwin Bell.

Drs. Henry C. Gunning, Thomas S. Lovering, and George Tunell reviewed the present manuscript and their constructive comments were very helpful.

We also appreciate the assistance received in assembling a representative collection of the Boulder ores. Dr. T. S. Lovering of the U.S. Geological Survey provided several polished sections, and numerous ore samples were donated by local mining men, particularly Fred Dopp, H. M. Williamson, Jr., Arch Walker, William Brewster, William Gibson, and Russel MacClellan. Other specimens were supplied by William Byron, a mineral dealer and former mining engineer in Boulder, and by Professor Russell Honea.

Electron probe microanalyses were performed at the Advanced Metals Research Laboratory under the direction of Dr. Sheldon Moll. Electron photomicrographs of telluride replicas were taken by Robert Fear and Robert Corbett under the direction of Professor Wilbur Bigelow of the Department of Chemical Metallurgy at The University of Michigan.

The chapters on geologic setting, structural relationships, and structural control have largely been drawn from U.S. Geological Survey Professional Papers 223 (Lovering and Goddard, 1950) and 245 (Lovering and Tweto, 1953), and from E. N. Goddard's experience in the telluride districts during his work with the U.S. Geological Survey in association with T. S. Lovering, and in later years.

Contents

ILLUSTRATIONS

PLATES

Plate   Page

2. a. Altaite (at)[1] and petzite (pz) in quartz (qz). Concentric aggregate of
   fine-grained gold in tellurium oxide (Au + tt) selectively replaced petzite.
   Reflected light, in air, 250 X. Horsefal vein; b. Altaite (at) veined by
   native gold (Au). Relict of sphalerite (sl) appears at edge of vug. Reflected
   light, in air, 183 X. Smuggler mine; c. Galena (gn) replaced by hessite
   (hs) which was in turn veined by native gold (Au). Reflected light, in air,
   218 X. Nancy mine; d. Tetrahedrite (td) interstitial to pyrite (py) has been
   replaced by native gold (Au). Reflected light, in air, 184 X. Colorado vein;

[1]Where possible, mineral abbreviations used in this report are those recommended by Chace
(1956).

FIGURES

TABLES

Abstract

The Jamestown, Gold Hill, and Magnolia mining districts of Boulder County, Colorado, provide one of the few classic examples of telluride mineralization in this country. These districts are parts of a broad, north-trending belt of telluride mineralization about 5 square miles in area located at the northeastern end of the Front Range mineral belt. The predominant country rocks are Precambrian granites, gneisses, and schists which are bounded on the east by Paleozoic and Mesozoic sediments upturned along the front of the range. The telluride veins represent one stage in a complex sequence of Early Tertiary ore types that show varying degrees of correlation with exposed Early Tertiary intrusives. Erosion has removed any volcanics erupted at the time of mineralization, but has exposed genetically related dikes and intrusion breccias of biotite latite in an area coextensive with that of the telluride mineralization.

A series of major northwest-trending faults called "breccia reefs" apparently served as the main channelways for introduction of the telluride ore fluids which rose from depth along the reefs and spread out into northeast-trending vein fissures where ore deposition took place. Most of the telluride production has come from only a few centers that show a pronounced structural control by the reefs. Local structural controls within the vein fissures include vein intersections, intersections of veins with earlier faults or with igneous dikes, irregularities of the veins as related to wall movements, and the physical character of the wall rocks. The telluride veins were mined over a collective vertical range of 3000 feet, and there is no clear-cut evidence of a bottom limit to the ores.

The telluride veins are typically composed of an interlacing network of pyritic or marcasitic horn quartz seams in which the ore minerals are quite sparse and irregularly distributed. Sixty-seven vein minerals have been identified (forty-one by X-ray methods). Individual polished sections commonly contain a dozen or more metallic minerals in fine-grained, intimate intergrowths filling or coating fractures or scattered vuggy openings in the fine-grained vein quartz. The chief ore minerals are sylvanite, petzite, hessite, and native gold and in some mines calaverite and krennerite are also important. Ten other tellurides, as well as native tellurium and a variety of

1

sulfides and sulfosalts formed in the telluride stage of mineralization but contributed little or nothing to the values. The principal gangue constituents are quartz and altered wall rock, but roscoelite, ankerite, calcite, fluorite, and barite are locally present.

A section of the report on problems of telluride identification gives data on the polarization figures, rotation properties, reflectivities, and indentation hardnesses of the tellurides.

Hypogene textures and associations of the telluride ores have in many cases been highly modified by cooling, but these effects, as well as the original depositional sequence itself, are clarified by experimental phase relations in the system Au-Ag-Te. In general, the original sequence was one of early sulfides, followed by native tellurium and a series of tellurides of progressively lower tellurium content, and finally by late native gold. A high degree of local equilibrium was maintained during the initial deposition, but as the ores cooled equilibration seems to have varied among assemblages of different bulk compositions. Certain telluride intergrowths formed upon cooling of unstable high temperature phases once present in the ores, and some of these changes took place long after the period of active mineralization as the mineralized terrane gradually cooled.

The individual telluride veins and the telluride belt as a whole are essentially unzoned. However, many of the separate productive centers have a distinctive mineralogy defined by unusual proportions or associations of minerals that are otherwise widespread in occurrence. These relationships are attributed primarily to variations in the bulk compositions of the ore fluids that mineralized the separate structural centers.

The telluride veins have not been deeply weathered and the residual enrichment of gold is correspondingly slight. Partially oxidized ore contains abundant jarosite, limonite, and tellurium oxides and in places some supergene tellurium, mercury, hessite, and the copper tellurides. Fine spongy gold in limonite ("rusty gold") is common in the outcrops and is in places associated with native silver and the silver halides. The geochemical behavior of the principal metals, gold, silver, tellurium, and iron is discussed in terms of acidities, oxidation potentials, and chloride ion activities in the oxide zone.

Based on physiographic evidence, the known telluride ores are estimated to have formed under a rock cover 2600 to 4600 feet thick and at confining pressures in the range 78 to 360 bars. This estimate is consistent with confining pressures indicated by the arsenopyrite "barometer." Numerous "thermometers" are applicable to the vein and wall rock assemblages and indicate depositional temperatures locally as high as 350°C, but generally in the range 250 to 100°C. At any point in the veins, depositional temperatures declined through time.

Tellurium is thought to have been transported along with the other cations as soluble chloride complexes in slowly moving ore fluids released from a biotite latite source underlying the telluride belt. These fluids may have acquired some or all of their Si, Fe, Ca, Mg, and possibly V and Ba from the altered wall rocks, but the other vein components including Te, S, and the precious metals were probably hydrothermal differentiates of the biotite latite.

A brief review of major telluride districts shows that there is no obvious scheme of genetic classification that can be based on tellurium mineralogy. Compared to other world districts, the Boulder County belt has produced ores of unusual variability, and the abundance of both free gold and free tellurium in a single major district is truly exceptional. The Boulder County deposits are best placed in the epithermal class of the traditional intensity scale and are an excellent example of complex Tertiary mineralization in Precambrian terrane.

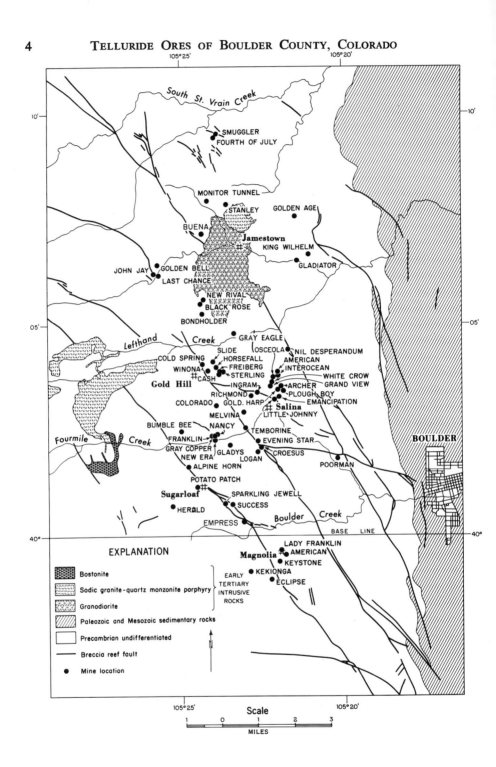

Figure 1. Telluride mines and prospects in Boulder County from which specimens have been collected for the present study.

Introduction

Boulder County, Colorado, has long been known as one of the few classical areas of telluride mineralization in this country. Its fame rests chiefly on the truly remarkable variety of rare metallic minerals that are abundant in the telluride veins. Some 35 different metallic species and a total of 67 vein-forming minerals have been identified; in individual ore samples it is not at all unusual to find ten or more minerals intergrown in such a complex fashion that identifications are impossible in hand specimen and difficult by any means.

Soon after the initial discoveries of gold telluride ore in this area in 1872, the unusual mineralogy of the veins drew the attention of such capable scientists as Genth (1874, 1877) and Silliman (1874a, 1874b) who soon established a descriptive catalogue of the principal telluride minerals. It is a tribute to these early workers that their careful record has required no major modifications in the light of more recent studies.

Much additional work has been done in connection with the telluride ores in the way of either detailed studies of selected mineral species (Wahlstrom, 1950), studies of specific districts within the general area of telluride mineralization (Goddard, 1935, 1940; Wilkerson, 1939; Lovering and Tweto, 1953), or general summaries of the geology and mineralogy of the area as a part of more comprehensive investigations of the entire Front Range mineral belt (Lovering and Goddard, 1950). To date, there has been no intensive investigation utilizing modern methods and concentrating specifically on the nature and origin of the telluride veins per se.

The present investigation was started in 1958, and has been primarily an X-ray and microscopic study of several hundred telluride ore samples from 53 mines and prospects throughout Boulder County (Fig. 1). About 200 polished sections were prepared of which 156 contained visible tellurides and were studied in detail. The abundances, associations, and paragenesis of the principal metallic minerals have been determined and interpreted in terms of the environment and processes leading to the formation of these unusual deposits.

History and Production

Gold was first discovered in Boulder County in 1859, when placer deposits were located near Gold Hill; although several thousand people rushed into the area, and numerous lodes were discovered, many miners were disappointed and mining declined after 1860. The discovery of gold telluride ore at the Red Cloud mine of the Gold Hill district in 1872, however, led to a tremendous revival of mining activity and by 1879 nearly all of the larger gold telluride mines had been located. Though the ore bodies were remarkably rich, most of them were small and erratically distributed so that mining was sporadic over the next sixty years. Another great impetus to gold mining in the area was given by the rise in price of gold in late 1933. Nearly all the larger mines were reopened and production steadily increased until May 1942 when practically all activity was terminated by the gold-mining restraining order of World War II. Since then, there has been little activity and the last of the significant telluride mines to be actively worked were the Poorman mine in 1953, and the Cash and Colorado-Rex in the late 1950's.

The total value of the telluride ore production from Boulder County can only be estimated because records kept by the U.S. Bureau of Mines (Lovering and Goddard, 1950, p. 12) do not distinguish between the various ore types. However, these records give the total value of precious and base metals produced from the county from 1858 to 1944 as $33,127,638 of which $24,395,593 was in gold and $8,019,291 in silver. On the basis of the production of some of the gold telluride mines, it is estimated that the telluride ores probably accounted for between 60 and 70 percent of the gold-silver total, or between $20 and $22 million. No significant production has come from the telluride mines since 1944.

Although tellurium was not recovered from the ores, a rough estimate of the total quantity extracted along with gold and silver can be made on the basis of these production figures and the mineralogical analyses presented in this report. Even if full allowance is made for native tellurium and tellurium in various tellurides other than those of silver and gold, the grand total removed from all of the veins in 94 years of sporadic mining is something less than 60 tons. This figure is less than half of the current annual production from domestic by-product sources alone and indicates that this erratic type of high-grade gold-silver mineralization has little significance as a future reserve of tellurium.

Geologic Setting

GENERAL FEATURES

The telluride ores of Boulder County are in a broad north-trending belt of about 50 square miles at the northeastern end of the Front Range mineral belt (Fig. 2). This area, herein referred to as the Boulder County telluride belt, includes the mining districts of Jamestown, Gold Hill, Magnolia, and Sugarloaf; some telluride ore was also mined in the small isolated district of

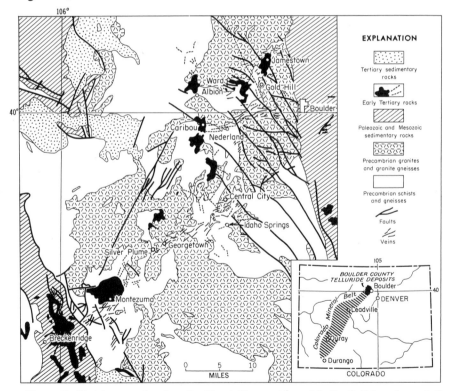

Figure 2. Sketch map showing the Front Range mineral belt of Colorado and the location of the Boulder County telluride deposits (after Lovering and Goddard, 1950, Fig. 1).

9

Eldora 10 miles southwest of the main area, and small amounts of telluride minerals have been reported from the Ward district about 4 miles to the west. However, the writers have few data on these latter ores, or on the relationships within the veins. They are therefore omitted from the discussions in this paper.

The entire telluride belt is within the Precambrian terrane of the Front Range, just west of the foothills belt of Pennsylvanian to Cretaceous sediments (Fig. 3). The principal Precambrian rocks are the schists and gneisses of the Idaho Springs Formation, small amounts of Swandyke Hornblende

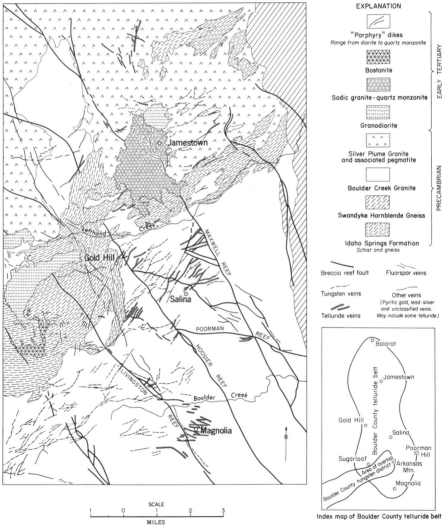

Figure 3. Geologic map of a part of Boulder County showing the distribution of telluride veins and of other ore types.

Gneiss and large bodies of the Boulder Creek and Silver Plume granites. These rocks have been intruded by Early Tertiary stocks and dikes, the so-called "porphyries" of the Colorado mineral belt, which are mainly monzonite but range in composition from diorite to sodic granite. A somewhat earlier diabase dike, the "Iron dike," extends northwesterly across the area, but apparently had little relationship to mineralization in the Boulder County mining districts. The telluride ores apparently bear a close genetic relationship to small, inconspicuous dikes of biotite latite, and biotite latite intrusion breccia intruded late in the porphyry sequence.

The telluride belt is cut by a system of strong northwest-trending faults, called breccia reefs by Lovering and Goddard (1950), and most of the profitable telluride ore shoots are in northeast-trending vein fissures within a few thousand feet of these strong faults (Fig. 3 and Fig. 7). In fact, most of the ore deposits in the Front Range mineral belt seem to have been localized by the intersection of breccia reef faults with the northeast-trending zone of porphyries and vein fissures.

PRECAMBRIAN ROCKS

Schists and gneisses of the Idaho Springs Formation form an irregular zone that trends northeast across the area and is caught between a batholith of Boulder Creek Granite on the south and one of Silver Plume Granite on the north. The chief rock types are quartz-biotite schist, quartz-biotite-sillimanite schist, and injection gneiss, but thin layers of quartzite and small lenses of lime-silicate rock are locally present. The general parallelism of the foliation and the vein fissures in these schists and gneisses resulted in tight, gougy seams with little width and only small amounts of ore. This is perhaps the reason for the relatively small extent of telluride mineralization in the Eldora district where the prevailing rock is schist.

Scattered bodies of hornblende-gneiss occur around the northern periphery of the telluride belt and small lenticular bodies of quartz monzonite gneiss are locally present, but these rocks contain no veins except for the John Jay mine.

The Boulder Creek Granite is the chief wall rock of the telluride veins, and the Gold Hill, Sugarloaf, and Magnolia districts are almost entirely within the northern part of a small oval-shaped batholith about 100 square miles in areal extent (Fig. 3). A few veins in the Jamestown district are at the northern edge of this batholith, and a few, notably the John Jay, are in the eastern edge of a large stock of this granite a few miles to the northwest of the batholith.

The batholith ranges in composition from quartz monzonite through granodiorite to granite, and in texture from medium-grained equigranular to coarse-grained porphyritic. The color is predominantly medium gray, but in the coarser varieties, large stubby pink microcline phenocrysts give a mottled pinkish-gray cast. The range in mineral composition as given by Lovering and Tweto (1953, p. 9) is "typically 20 to 40 percent quartz, about 20 percent microcline, 25 to 30 percent plagioclase (oligoclase or andesine), and 15 to 35 percent biotite." In most of the batholith, the parallel orientation of the coarse-grained biotite gives a distinct to pronounced gneissic structure which is parallel to the borders and is believed to be primary. The coarse-grained porphyritic granite is chiefly in the Gold Hill district in the northern part of the batholith and studies of the foliation and lineation seem to indicate that magma rose from a deep-seated conduit under this area and fanned out southward and upward (Lovering and Tweto, 1953, p. 13, Fig. 10).

Along its borders, the batholith is strongly gneissic, and interfingers with or locally grades into granitized schists. Vein fissures parallel to the foliation tend to be tight and gougy, whereas those parts that cut across the gneissic structure were more favorable for ore deposits. The coarse-grained parts of the batholith constitute the most favorable host rock.

Silver Plume Granite is the host rock for most of the telluride veins in the Jamestown district. These deposits are in the southern part of a large compound batholith made up of numerous coalescing stocklike bodies with wedge-shaped septae of schist and gneiss caught in between. This granite is a light- to pinkish-gray, medium- to coarse-grained, slightly porphyritic rock in which lath-shaped, closely packed orthoclase phenocrysts form a well-developed primary flow structure. The groundmass is made up of quartz, oligoclase, biotite, and muscovite. The texture and composition of this granite make it a very favorable host rock, and trends of vein fissures parallel to or crosscutting the flow structure seem equally favorable sites for ore deposition.

Dikes and irregular bodies of pegmatite and aplite are common throughout the telluride belt in both types of granite and in the schists and gneisses; however, few ore bodies are found in the pegmatites and aplites. Most veins are barren within pegmatite bodies, but many telluride veins in the Gold Hill district locally follow the walls of pegmatite dikes. Even here, however, the ore bodies are in brecciated granite adjacent to the pegmatite. Medium- to fine-grained aplite dikes do not appear so favorable as the adjacent coarser-grained granite.

LATE CRETACEOUS-EARLY TERTIARY IGNEOUS ROCKS

General Relations

The Precambrian rocks of Boulder County have been intruded by a sequence of Late Cretaceous-Early Tertiary stocks and dikes, the so-called "porphyries" of the Front Range mineral belt (Fig. 2). These rocks have been discussed in detail by Lovering and Goddard (1938), who inferred their origin by differentiation from a deep-seated basaltic magma and recognized a close genetic relationship between various mineral deposits and certain rock types in the sequence. The telluride ores of Boulder County seem closely related to the latest rocks of this sequence, the biotite latites and associated intrusion breccias.

Rock Types

These "porphyries" range in composition from diabase to sodic granite. Most of them are porphyritic and in nearly all, the groundmass is fine-grained to crypto-crystalline. The various rock types, their sequence of intrusion, and the various ore deposits that seem to be genetically related to them are given in Table 1.

The diabasic gabbro forms a strongly persistent dike, called the "Iron dike," that trends about N. 25° W. along the western border of the telluride belt and can be traced in several segments for nearly 30 miles through the Precambrian terrane. No ore deposits appear to be genetically related to this rock.

TABLE 1. DISTRIBUTION AND INTRUSIVE ASSOCIATIONS OF PRINCIPAL ORE TYPES
[COMPILED FROM LOVERING AND GODDARD (1950) AND LOVERING AND TWETO (1953)]

Type of Deposit	Associated Intrusives	Chief Minerals	Distribution and Remarks
TUNGSTEN		Ferberite, horn quartz, a little pyrite	Youngest ore type. Occurs in Boulder County tungsten district, which extends 9½ miles west-southwest from Arkansas Mountain (see Fig. 3). Chief production and apparent source area in western part of this district
GOLD-SILVER TELLURIDE	Small dikes and irregular bodies of biotite latite and intrusion breccia	Gold-silver and associated tellurides, native gold and tellurium with accessory sulfides in pyritic horn quartz	Roughly elliptical 50 square mile area encompassing Jamestown, Gold Hill, and Magnolia districts and satellite Eldora district 3 to 6 miles southwest of Nederland. Gold Hill most productive district
PYRITIC GOLD	Dikes of bostonite	Pyrite, quartz, free gold, chalcopyrite, roscoelite	Widespread. Important veins in Jamestown and Gold Hill districts but also noncommercial veins in Magnolia and Tungsten districts. Some pyritic gold ores of the Gold Hill district are younger than the tellurides
FLUORITE	Stocks and dikes of sodic granite and quartz monzonite	Fluorite, quartz, sericite and clay minerals with included fragments of lead-silver ore	Concentrated in zone ½ x 2 miles along the southwestern margin of Porphyry Mountain stock in Jamestown district
LEAD SILVER		Argentiferous galena, tetrahedrite and lead-silver sulfantimonides with sphalerite chalcopyrite and minor free gold. Pyrite and quartz	Oldest ore type. Spotty distribution. Important veins and pipe-like bodies in main fluorite zone at Jamestown and at several locations in Gold Hill district. Some lead-silver ore is late and apparently post-dates the tellurides

Stocks of granodiorite and alkalic granite-quartz monzonite are a prominent feature of the Jamestown district (Fig. 3), and irregular stocks of alkalic syenite and bostonite occur just west of the Gold Hill district. Dikes of hornblende diorite, biotite-quartz monzonite, rhyolite, alkalic quartz monzonite, and bostonite are scattered throughout most of the telluride belt. Most of them are from a few feet to 30 feet wide and have a northeasterly trend, though some trend east or northwest.

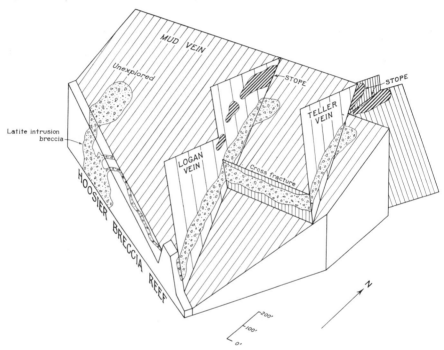

Figure 4. Block diagram showing the relation of the biotite latite intrusion breccia to gold telluride ore shoots in the Logan mine (from Lovering and Goddard, 1950, Fig. 24).

Small, mostly inconspicuous dikes of biotite latite and latite porphyry and biotite latite intrusion breccia deserve special mention because of their apparent close genetic relationship to the telluride ores. The biotite latite porphyries are dense, dark-gray rocks containing small phenocrysts of biotite, hornblende, and plagioclase (chiefly andesine) in a microcrystalline groundmass of oligoclase and orthoclase. Locally the groundmass is glassy. These dikes are sparingly scattered in the northern, central, and southwestern parts of the Boulder County telluride belt (see Fig. 7). A fairly large dike several feet wide lies between the Red Cloud and Cold Spring telluride veins. Small dikes occur just south of Wallstreet and in the Logan and Yellow Pine mines at the southwestern edge of the Gold Hill district. In underground workings along the Logan and Yellow Pine veins, and along the

Twin vein in the northern part of the district, thin seams of biotite latite intrusion breccia are exposed. This rock has a soft gray gouge-like appearance and is crowded with small granite and pegmatite fragments. Minute biotite phenocrysts are visible in places. The groundmass is made up of microcrystalline laths of feldspar in glass. Lovering (*in* Lovering and Goddard, 1938, p. 64) was the first to recognize the genetic association of these rocks with the telluride ores, and illustrated this relationship in a block diagram of veins in the Yellow Pine and Logan mines (see Fig. 4).

STRUCTURE

Most of the significant structural features of the Boulder County telluride belt were formed during the Laramide orogeny, though some Laramide fault trends were influenced by Precambrian shear zones (Goddard, 1940, pp. 115–116; Tweto and Sims, 1963, pp. 991–1014). The broad anticlinal uplift of the Front Range in Laramide time occurred in almost the same position as the earlier Ancestral Rockies uplift in Pennsylvanian-Permian time. Normal, steep reverse and thrust faults flank the range in many places and strong northwest-trending normal and strike-slip faults, called breccia reefs by Lovering and Goddard (1950), cut diagonally across the range on the east side of the divide.

Precambrian Shear Zones

Several of the Laramide faults and vein fissures have broken in part along planes of weakness marked by Precambrian shear zones in the Boulder Creek granite. Notable examples are the Poorman breccia reef, and the Slide, Cash, Richmond, and Grandview veins. The shear zones range from 5 to 200 feet wide and consist chiefly of sheared but rather fresh-looking granite which resembles strongly foliated gneiss. Commonly there is a faint greenish tint due to chloritic alteration of biotite and locally there are thin stringers of epidote. Apparently this shearing took place shortly after the intrusion of the Boulder Creek granite as some pegmatites were involved in the shearing while others were intruded later.

Breccia Reef Faults

Strong, extensive breccia reef faults are the most prominent structural features in the Boulder County telluride belt and exerted a most significant influence on the distribution of the telluride ores. The local miners were quick to recognize the spatial relationships of the telluride and other ore bodies to these large faults, which they called "dikes" because in many places the silicified fault zones form prominent dike-like outcrops. In fact nearly all the principal breccia reefs in Boulder County were identified and named by the miners.

The telluride belt is cut by three of the northwest-trending breccia reefs, the Maxwell, Hoosier, and Livingston, which are nearly parallel, spaced 1½ to 2½ miles apart and can be traced for 15 to 20 miles (Fig. 7). They range in strike from N. 20° W. to N. 70° W. and all are steeply dipping. At the northeastern margin of the telluride belt, the Standard reef also strikes northwest, but dips only 15° to 20° SW. Two other significant breccia reefs, the Poorman and Fortune, have a more westerly trend and cut across from the Maxwell to the Hoosier. The Poorman is nearly vertical, but the Fortune dips from 30° to 35° N. Two other smaller breccia reefs, the Blue "vein" and the Bull of the Woods, are N. 70° W. branches of the Maxwell and some of the important telluride veins terminate against them.

The breccia reef faults range in width from a few feet to about 300 feet, and consist of brecciated and sheared zones which are gougy in places and locally expand into wide sheeted zones. In many places these fault zones are silicified and contain finely disseminated hematite, which gives the zone a purplish, reddish, or pinkish color. In some places, this coloration extends out into the wall rock on either side. Locally, as for example in the Hoosier reef a mile southeast of Gold Hill (Fig. 3), milky "bull quartz" veins, a few feet to 30 feet wide, occupy the fault zone and form prominent dike-like outcrops. The silicification and red hematite coloration are a great aid in tracing these faults through the Precambrian terrane. However, where not silicified, the faults are difficult to trace, and are characterized by ill-defined hematitic shear zones partly chloritized in places. The Fortune reef, which is only a few feet wide in most places, is largely unsilicified and thus is difficult to trace on the surface, though the red hematite coloration is commonly present. In a few places breccia reefs contain small amounts of pyrite, chalcopyrite, and fluorite, but nowhere are these abundant enough to be commercial.

Vein Fissures

Nearly all the veins of the telluride belt fill fault fissures that are younger than the chief movement and mineralization along the breccia reefs. Most of the veins strike northeast almost at right angles to the reefs, but notable exceptions are those of the Poorman group which strike nearly east, parallel to the Poorman breccia reef, and in the Magnolia district where the principal trends are either west or west-northwest (Fig. 3). Nearly all the veins dip steeply and the predominant dip of the northeasterly veins is to the southeast.

The vein fissures are of relatively small extent; few can be traced on the surface for more than 1½ miles and some of the most productive, as for example the Buena, are less than half a mile long. The vein fissures are mostly narrow and the zone of brecciated, fractured and sheared rock commonly ranges from a few inches to 15 feet in width. Displacements on the vein fissures are small, commonly a few feet, but range from a few inches to as much as 30 feet. On most of the northeast-trending veins the southeast wall has moved southwest and down at a low angle, but on the veins of westerly trend, the north wall has moved west and down at a low angle. As will be discussed later, this movement related to irregularities in the veins has had a significant control on the distribution of the telluride ore bodies.

Ore Deposits

GENERAL REMARKS

Although the gold telluride ores have accounted for most of the gold and silver produced from the telluride belt of Boulder County, a number of other vein types have been mined and significant amounts of other ores have been produced. Lead-silver and pyritic gold veins in the Gold Hill and Jamestown districts have contributed appreciable amounts of gold, silver, lead and zinc. Tungsten has been mined in the area of overlap of the Gold Hill district and the Boulder County tungsten district (see southwestern part of Fig. 3), and also in the Magnolia district. Large tonnages of fluorspar have been produced from veins and breccia zones in the Jamestown district, and a few shipments of nickel ore came from a small mine at the western edge of the Gold Hill district (Goddard and Lovering, 1942). The distribution of these various vein types is shown in Figure 3, but it has been necessary to lump the lead-silver and pyritic gold veins because in many places these veins contain both types of ore and it is difficult to distinguish between them. The various ore types are discussed briefly below, except for the nickel deposit of Precambrian age at Gold Hill. All the Early Tertiary ores appear to have a broad genetic relationship, and some of the mineral assemblages are early fillings in the telluride veins. Also, some of the earlier veins have been effective in the structural control of telluride ores in the later veins.

LEAD-SILVER DEPOSITS

Lead-silver deposits are widely scattered in the telluride belt, but most of the production has come from three small areas: (1) on the west side of the Porphyry Mountain stock, 1½ miles northwest of Jamestown, (2) in the vicinity of Summerville, just southeast of the junction of Hoosier and Fortune reefs, and (3) in the vicinity of the junction of the Hoosier and Poorman reefs. Most of the deposits are in veins, but in the Jamestown area breccia fragments of lead-silver ore are abundant in parts of some fluorspar breccia zones, and in the Alice mine a small pipe-like body of lead-silver ore occurred on the footwall of an irregular fluorspar vein. The breccia fragments, many of which contain abundant galena, appear to represent early veins broken

17

up by the brecciation that preceded the fluorspar mineralization. Most of the lead-silver veins strike northeast, but some trend northwest and two of this group are along the wall of the Hoosier breccia reef.

The chief ore minerals of the lead-silver veins are argentiferous galena and gray copper (both tennantite and tetrahedrite) with variable amounts of sphalerite, chalcopyrite, and pyrite. In a few mines, freibergite and stromeyerite occur with the gray coppers and add to the silver values. Lovering and Tweto (1953, pp. 52, 39–41, 185) have described complex silver sulfide and sulfosalt assemblages present in pre-tungsten lead-silver ores of the Yellow Pine mine and the same assemblages were noted in ores of the Little Johnny, Grandview, and Croesus veins in the present study. Glassy to milky quartz is the chief gangue mineral but in a few lead-silver veins in the southern part of the Jamestown district the gangue is horn quartz with late ankerite and some barite. These particular veins may be younger than the telluride deposits, but as a general rule the lead-silver deposits are the oldest of the Early Tertiary mineralization in the telluride belt.

Most of the lead-silver ore shipped contained from .06 to 1.50 ounces of gold per ton, 2.8 to 300 ounces of silver per ton, 1 to 40 percent lead and 0 to 5 percent copper. Rarely was zinc present in commercial quantities.

FLUORSPAR DEPOSITS

The fluorspar deposits are confined to the Jamestown district where they occur in a northwest-trending belt about two miles long and one mile wide on the south and southwest sides of the sodic granite-quartz monzonite stock in the vicinity of Jamestown (Fig. 3). Most of the production has come from a small area of about a quarter of a square mile on the west side of the stock. A detailed description of these deposits as explored up to 1945 is given by Goddard (1946). Fluorspar production began with 400 tons in 1903 and has been fairly continuous up to the present time, with a total production of several hundred thousand tons. Most of the production up to 1940 was metallurgical "spar," but in more recent years the chief product has been acid grade spar for the chemical industry.

The fluorite[2] occurs in relatively short breccia zones and veins of predominant northwest and northeast trend, though some strike north and a few others nearly east (Fig. 3). Nearly all are steeply dipping. The breccia zones are large lenticular bodies of brecciated granite and fluorite cemented by a fine-grained mixture of fluorite and clay minerals. Scattered fragments of lead-silver ore and coarse milky quartz are irregularly scattered through some of the zones.

The fluorite veins range from a few inches to 20 feet in width, and from 150 to 1000 feet long. They also are filled with brecciated fluorite, but contain

---

[2]Fluorspar is the commercial name and has been so used up to this point. From here on the discussion is largely mineralogical so the mineral name "fluorite" will be used where appropriate.

relatively few fragments of granite or granodiorite. The two most productive mines, the Burlington and the Emmett, are on combination breccia zones and veins, and have both been explored to a depth of more than 1000 feet. In the lower levels the ore bodies are still strong, but are somewhat lower grade and seem to be consolidating into veins with less breccia than above.

Most of the fluorite in the district is deep violet in color, some almost black, but various shades of purple are common and some is white to greenish white. In the Emmett mine and locally in the Burlington, coarse-grained white to purple fluorspar is cemented with the fine-grained deep violet variety. The deep violet color appears to be due to irregularly scattered minute grains of pitchblende, which can be seen under the microscope surrounded by deep purple halos. The deepest-colored fluorite shows as much as .0515 percent uranium equivalent (Goddard, 1946, p. 19).

A great variety of sulfide minerals is irregularly scattered through the fluorspar deposits, many intergrown, but mostly occurring as fragments in the breccia. Thus they seem to be largely associated with an earlier period of lead-silver mineralization. The mixed sulfides include galena, sphalerite, chalcopyrite, pyrite, tennantite, and enargite. Appreciable silver and small amounts of gold are associated with these sulfides. Shipments of lead-silver ore sorted from the fluorspar breccia zone of the Argo mine during the period 1900-1933 contained 0.06 to 0.3 ounces of gold and 7 to 26 ounces of silver to the ton, 22 to 40 percent lead and as much as 5 percent copper. The silver seems to be associated with the galena and tennantite, but the source of the gold is unknown. In the west end of the Chancellor fluorite vein, a small pipe-like body of gold ore was mined to a depth of 120 feet below the surface. This averaged about 4 ounces of gold and 5 ounces of silver to the ton. It was reported by some of the miners that this was gold telluride ore, but the writers were unable to verify this.

PYRITIC GOLD DEPOSITS

Pyritic gold ores occur in northeast-trending veins in various parts of the telluride belt. Most of them are low grade, and in many veins that have had little production, it is difficult to tell a pyritic gold from a telluride vein, particularly where all that shows on the dump is a little horn quartz and disseminated pyrite. Accordingly, it is quite probable that some telluride-type veins may have been lumped with the pyritic gold veins in Figure 3.

The typical filling of the pyritic gold veins is coarse-grained pyrite intergrown with some chalcopyrite in a gangue of sugary to glassy quartz which grades into horn quartz in places. Small amounts of sphalerite, galena, and gray copper are also common. Rich pockets of free gold have been found in a few of these veins, notably the Klondike northeast of Gold Hill, the Logan at the junction of the Hoosier and Poorman reefs, and the New Era of the Wood Mountain group. However, in most of the pyritic gold veins, free gold is not visible; the best values seem to be associated with the chalcopyrite, but low-grade values occur in the pyrite. Ankerite is common in some of the veins, where it is the latest mineral present. In the Grand Republic mine at the junction of the Hoosier and Poorman reefs, a zone of interlacing pyritic horn quartz seams and finely disseminated pyrite formed a low-grade ore body that was mined successfully for several years. It was suggested (Guiteras, 1937) that the gold values were in tellurides, but so far as is known no telluride minerals were identified.

Although there are other cases where it is difficult to distinguish between pyritic gold and telluride veins, in many places the pyritic gold ores seem definitely earlier. Wherever horn quartz occurs in pyritic gold veins, it is clearly younger than the intergrowth of coarse pyrite and glassy quartz. Bismuth telluride occurs in pyritic gold ore of the Stanley mine on the west side of the Porphyry Mountain stock, but here the assemblage is bismuth-inite-tetradymite-gold and occurs in vugs in coarse intergrowths of pyrite and milky to glassy quartz. This assemblage has not been found elsewhere in typical telluride ores.

GOLD TELLURIDE DEPOSITS

Most of the gold telluride veins occupy northeast-trending fissures in the telluride belt and their distribution is shown in Figure 3. Although many of these veins can be traced for a distance of from 1 to 1.5 miles, only relatively small parts of them have been productive and most of the values have been confined to the centers of mineralization shown in Figure 7. These veins commonly range in width from a few inches to five feet, but in a few places are as much as 15 feet wide. Most of the productive veins have been explored to depths ranging from 300 to 700 feet below the surface (Fig. 12). The two deepest mines are the Slide, at 1080 feet, and the Ingram, at 950 feet vertical depth. In most of the mines the veins have not pinched out in the bottom levels, but commonly water problems and other mining difficulties have made it unprofitable to explore deeper.

The telluride veins characteristically consist of interlacing dark-gray to nearly white horn quartz seams enclosing breccia fragments and lenticular bodies of sheared and altered wall rock, chiefly granite (Pl. 1). These seams generally range in width from ⅛ inch to 18 inches. There are a few veins 2 to 3 feet wide of solid horn quartz, but these are mostly barren. Locally, at vein junctions and intersections, much wider zones have received telluride mineralization. For example, in the Buena mine near Jamestown, at the junction of the Buena vein with the Grant and Central Pacific, the "Big Stope" ore body consisted of a highly shattered zone or stockworks 30 feet wide; other similar ore bodies in that mine ranged from 10 to 15 feet wide.

In some mines, thin veinlets of horn quartz and telluride minerals extend out into the wall rocks for a few feet and are mined as part of the ore body.

The telluride minerals occur in groups of blades or irregular small masses in the horn quartz seams (Pls. 8e, 9h and 10c) and in many places these minerals are so complexly intergrown that they cannot be easily identified microscopically. Petzite, hessite, sylvanite, and native gold are the chief ore minerals, though calaverite and krennerite are significant in some mines. Ten other telluride minerals and native tellurium have been identified in the ores, but have contributed little or nothing to the value. Finely disseminated pyrite and marcasite are nearly everywhere associated with the tellurides in the horn quartz, and pyrite is commonly disseminated in the wall rock. Very small amounts of galena and sphalerite are irregularly distributed and locally other sulfides and sulfosalts are present. In some ore bodies significant amounts of free gold are intergrown with the telluride minerals and contribute much to their value. The horn quartz and altered granite are the chief gangue minerals, but locally, ankerite-dolomite, calcite, roscoelite, barite, and fluorite are present. The mineralogy, textures, paragenesis, and origin of these various ore and gangue minerals as well as the grade of ore are discussed in later chapters.

The telluride ore bodies are generally small and discontinuous, so that when one ore body is worked out, considerable expense is necessary in searching for others. In most of the larger mines, the ore bodies have ranged from 50 to 400 feet in length, from 30 to 300 feet in breadth, and from a few to 15 feet in width. However, in the Slide vein a single ore body 60 to 230 feet in breadth was mined from the surface to a depth of 1000 feet. The "Big Stope" ore body in the Buena mine had a length of 150 feet, a breadth of 60 feet, and a width of 30 feet. Most of the larger ore bodies rake steeply and some are nearly vertical (see Fig. 5). In many of the mines, local pockets of unusually high-grade ore ranged from 10 to 30 feet in length, 2 to 20 feet in breadth, and a few inches to 2 feet in width. For the most part, these high-grade pockets are parts of lower-grade, larger ore bodies. It has been common practice among the miners for years to sort and sack the high-grade ore in the stopes and to ship the medium- and low-grade ore in car-load or truck-load lots.

TUNGSTEN DEPOSITS

The tungsten ores of Boulder County are in an east-northeast-trending district largely separate and distinct from the telluride belt, but in the area of overlap (Fig. 3) some tungsten mineralization was found in telluride veins, and in a few mines, notably the Logan, Kekionga, and Red Signe, small tungsten ore bodies were mined. The tungsten district, including the area of overlap, is discussed in detail by Lovering and Tweto (1953), and the following discussion is taken from their report.

In the tungsten district and in the area of overlap, the chief tungsten ore mineral is ferberite, though in a few places enough manganese is present to class the mineral as low-manganese wolframite. In most of the veins the ferberite forms the matrix of a breccia of horn quartz and country rock, but in some mines there are solid seams of high-grade ferberite and locally there are vuggy parts of veins containing ferberite crystals. Most of the ferberite is finely intergrown with variable amounts of quartz ranging from 1 or 2 percent to mostly horn quartz colored by fine ferberite.

The average width of the tungsten veins mined was 6 to 12 inches, but some high-grade shoots ranged from 2 to 5 feet. The ore is notably spotty in its occurrence, and no change in the character of the ore was noted from the surface to the deepest parts of the workings other than an abrupt lowering in grade at the bottom of the shoots. However, changes in wall rock alteration with depth were noted in places.

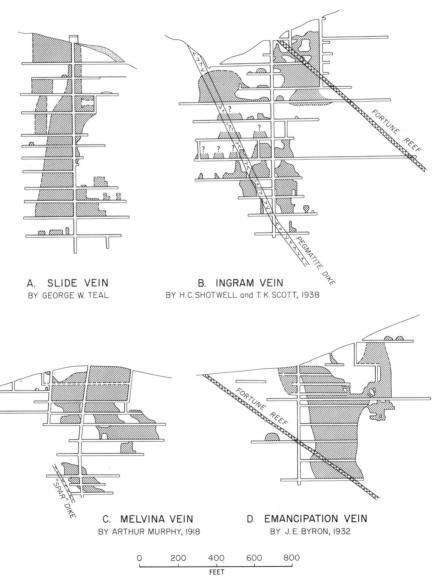

A. SLIDE VEIN
BY GEORGE W. TEAL

B. INGRAM VEIN
BY H.C. SHOTWELL and T.K. SCOTT, 1938

C. MELVINA VEIN
BY ARTHUR MURPHY, 1918

D. EMANCIPATION VEIN
BY J.E. BYRON, 1932

0   200   400   600   800
FEET

Figure 5. Stope maps of some of the principal telluride mines of Boulder County, Colorado.

In the zone of overlap, in the southwestern part of the telluride belt, the small tungsten ore shoots were mostly separate from the telluride ore bodies, but some specimens show both telluride minerals and ferberite in close association. On the basis of these specimens, Lovering and Tweto (1953) came to the conclusion that most of the tungsten ore was deposited later than the tellurides, but that locally small amounts of ferberite were earlier. The details of these relationships are discussed in later pages of this report.

AGE RELATIONS OF ORE TYPES

The age relations of the various types of ores discussed above have been worked out by Lovering and Goddard (1950) for the Front Range mineral belt as a whole, and seem to hold fairly consistently for Boulder County. These age relationships and those of the Tertiary intrusive rocks (the so-called porphyries) which appear to be genetically related are shown in Table 1.

In the Jamestown district, the age relations seem to fit fairly closely into the regional picture. The chief lead-silver ores are definitely the earliest, followed in sequence by the fluorspar, pyritic gold, and telluride ores. The zonal arrangement of these deposits around the Porphyry Mountain stock has already been mentioned. At the outer edge of the telluride zone a few minor lead-silver veins and tungsten-bearing veins appear to be later.

In the Gold Hill district, the age relations are not as clear. Most of the lead-silver ores seem to have followed closely the breccia reef period of mineralization and therefore are thought to be early. Much of the pyritic gold ore appears to be younger than the lead-silver ore and older than the telluride ore, but some of the pyritic gold ore is last, as indicated by a few places in the Slide vein where veinlets of typical pyritic gold ore cut horn quartz seams containing gold tellurides (Goddard, 1940, p. 125). In the area of overlap of the telluride belt and the tungsten belt, Lovering and Tweto (1953, p. 54) found evidence that some of the tungsten mineralization is earlier than the telluride ore, but they believe that most of the evidence indicates that the bulk of the tungsten ore is later.

Thus, though the age relations in the telluride belt seem to fit in general with those of the Front Range mineral belt as a whole, there is some conflicting evidence, and it appears that there is some alternation or overlap of the different ore types.

Wall Rock Alteration Along the Telluride Veins

There have been several studies of hydrothermal wall rock alteration associated with the different ore deposits of Boulder County (Lovering, 1941; Tweto, 1947; Lovering and Goddard, 1950; Lovering and Tweto, 1953; Bonorino, 1959), and a brief review of the subject is presented in the following paragraphs with emphasis on data that pertain to the telluride stage of mineralization. The present authors have not undertaken a detailed alteration study in the laboratory and so their contribution is confined to field and underground observations on the general character of alteration along the telluride veins.

The most detailed alteration study was a thin-section analysis of altered Boulder Creek "granite" bordering the tungsten veins, conducted by Lovering (1941) and Lovering and Tweto (1953). They were concerned primarily with the tungsten deposits, but did extend their observations to the telluride veins which they believe are closely related to the tungsten ores. They recognized two principal alteration zones in the tungsten district, an inner sericitic casing up to several feet wide and an outer zone of complex argillic alteration which locally extends up to 50 feet from the veins. In the inner zone, which is typically thinner than the vein, the chief alteration is a replacement of oligoclase by fine-grained dickite, hydromica, and sericite, which also replace one another in the order named. Nearly all biotite in this zone was converted to sericite, hydrous mica, or chlorite, while ilmenite was altered to pyrite or magnetite associated with secondary sphene and leucoxene. Original quartz, K-feldspars, and apatite show little alteration but are cut by veinlets of secondary quartz, sericite and hydromica. Wholesale silicification was observed in rock close to the veins in a few places but, in general, silicification was of minor importance in the tungsten district. In the argillic zone, which is normally 5 to 20 times the width of the vein, Lovering and Tweto distinguished three complex subzones with the outer zones carrying the earlier products of the alteration sequence. These subzones include an inner dickite zone next to the sericitic casing, an intermediate beidellite zone, and an outer zone of mixed allophane, hydrous mica, and sericite which grades imperceptibly into fresh rock. Lovering and Tweto (1953)

25

note that a little sericite actually occurs throughout the entire alteration envelope but is abundant only in the inner sericitic casing next to the veins. Plagioclase was most strongly attacked in each of the argillic subzones, but ilmenite was also replaced by hematite, leucoxene, and carbonate. Biotite was moderately stable, and original quartz, orthoclase, and microcline remained fresh in the argillic zone. The complex paragenesis of alteration products in the argillic subzones described by Lovering and Tweto (1953, pp. 59-60) cannot be repeated here, but it is worth stressing that these workers found definite textural evidence of a sequence with progressively younger products in zones closer to the veins.

With brief reference to the telluride veins, Lovering and Tweto (1953, pp. 58-59) state that the wall rocks are typically sericitized, silicified, and pyritized, but that a distinct outer argillic zone is seldom present. This is also the general impression of the present writers though in our experience an outer argillic zone is not uncommon particularly along telluride veins of the Gold Hill district. Where present, the argillic alteration may extend from inches up to about 10 feet from the sericitic casing and is most extensively developed along vein junctions or the intersection of veins with the breccia reefs. Wall rock caught up in the networks of interlacing horn quartz seams that comprise the veins is moderately to strongly silicified and the silicification in places extends several inches into the bordering wall rock. The predominant sericitic alteration extends from several inches to several feet from the veins and is commonly strongest close to ore bodies. As previously noted by Lovering and Goddard (1950, p. 265), disseminations of fine-grained pyrite are as a rule most abundant in the altered rocks close to ore but exceptions do occur.

Bonorino (1959) has described six different alteration patterns which he regards as representative of different areas or different ore types, or both, within the telluride and tungsten districts. Three of these patterns seem to apply, though not exclusively, to telluride veins. His Pattern III is allegedly representative of the entire Jamestown district, though none of the lead-silver veins were studied, and no clear distinction is made between the gold telluride and older pyritic gold stages of mineralization. This pattern consists of a single zone of predominant hydromica and secondary orthoclase replacing oligoclase and this is superposed on a mild greisen-like alteration which is widespread in the Jamestown area and reportedly associated with the early fluorite stage of mineralization. Bonorino (1959, p. 69) notes that there are veins in the Jamestown district that are bordered by a zone of "exclusively hydromica (or sericite) alteration" and cites the Grand Central telluride mine as a specific example. The most complex alteration observed by Bonorino (Pattern VI) consists of four zones which are, from the vein outward: the hydromica, secondary orthoclase, kaolinite, and montmorillonite zones. He indicates that this pattern is found in "the Nederland tungsten district and the Gold Hill pyritic gold district" (Bonorino, 1959, p. 59), but from the list of mines considered it is clear that a number of telluride veins are included in this category. Pattern V is equivalent to VI except that the zone of secondary K-feldspar is missing. This pattern was found in two small areas of the Gold Hill district and is evidently based upon study of a pyritic gold vein and a tungsten vein. Bonorino correlates this pattern with the zoning described for the tungsten district by Lovering and Tweto (1953), but there are significant differences in the mineral identifications that probably accrue in part from more refined X-ray methods used by Bonorino. For examples, Bonorino finds that the main alteration mineral in the "sericitic" casing is actually hydromica, and that kaolinite is the only representative of its group in the altered wall rocks. Contrary to the findings of Lovering and Tweto (1953), Bonorino (1959, p. 76) observed no evidence of mutual replacement among the hydrothermal alteration products and came to the conclusion that the different alteration zones formed simultaneously without the encroachment of one zone upon another.

The altered wall rocks bordering the telluride veins have never been analyzed chemically, but in view of their mineralogical similarity to the tungsten ore envelopes, the chemical exchanges are probably the same or similar to those known for the tungsten district (analyses of Bonorino, 1959; Lovering and Tweto, 1953). As noted by Bonorino (1959, p. 53), the total matter transferred during alteration was small, although the effects are mineralogically conspicuous. The principal trends noted by Bonorino were additions of K_2O, Al_2O_3, and H_2O to the wall rocks and removal of CaO, Na_2O, total iron, and SiO_2. By recomputing the analytical data of Lovering and Tweto (1953) to show gains and losses in terms of gram equivalents, Hemley and Jones (1964, p. 545) have demonstrated the importance of hydrogen metasomatism as a factor in part controlling alteration in both the sericitic and argillic envelopes.

The present authors feel that much additional field and laboratory study would be required with emphasis on the multiple stages of mineralization before a sound theory of alteration can be offered for the telluride ores in Boulder County. Such a study would be a major undertaking that should involve not only an analysis of hydrothermal events that produced the various ore types but an investigation of the early regional alteration and the more local effects along the system of breccia reefs. Some of the features already known, such as the curious zone of secondary K-feldspar (with kaolinite), are difficult to explain in terms of experimental equilibrium relations (Hemley and Jones, pp. 558–563; Fournier in Roedder, 1965, pp. 1392–1393) and are certainly worthy of additional research.

Structural Control of the Telluride Ores

REGIONAL CONTROLS

Although the telluride deposits are widely scattered over a broad north-trending belt, most of the production has come from only a few concentrated centers which exhibit a strong structural control. These centers of telluride mineralization are shown in Figure 7 and have been named after a prominent mine in each locality, with the exception of the Magnolia center which comprises the entire district. In the Gold Hill district, most of the centers include production from several veins, but in the Jamestown district a single vein in each center has generally accounted for nearly all the production as, for example, the John Jay, Buena, and Smuggler. As shown in Figure 7, most of these centers are along or close to one of the prominent northwest-trending breccia reefs, and some of the most productive centers are at the junctions of two breccia reefs as, for example, the Poorman center at the Maxwell-Poorman reef junction and the Logan center at the Hoosier-Poorman reef junction (see Fig. 7).

It is also worthy of note that virtually all the telluride production has come from ore shoots less than 2000 feet from a breccia reef. The only conspicuous exception is the Black Rose center, which is about one mile from the Hoosier reef. This striking spatial relationship of the telluride ores to the breccia reefs seems to indicate, as suggested by Lovering and Goddard (1950), that the breccia reefs served as trunk channels along which the ore solutions rose from depth and spread out into the more open vein fissures. The tightness of the reefs, due to strong shearing, silicification, and deposition of early bull quartz, left little space for telluride mineralization along these fissures, and only locally are small amounts of telluride ore found along them. However, the northeast-trending vein fissures, which probably opened at the critical time with small displacements and brecciation but relatively little tight shearing or strong crushing, provided open spaces for the ore deposition. Presumably, precipitation was favored by a drop in pressure as the ore fluids flowed from the tight reef structures into these relatively open channels.

In several places, some parts of the breccia reefs appear to have served

29

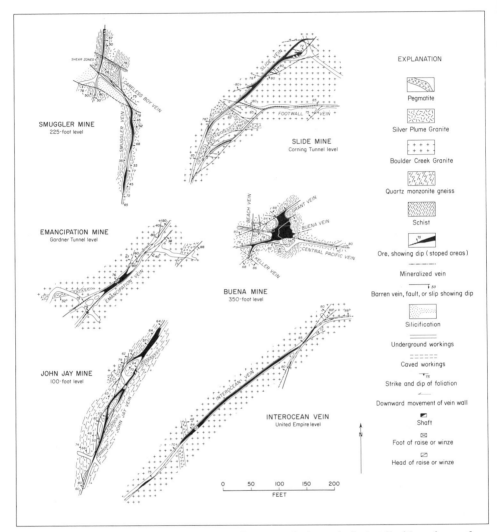

Figure 6. Geologic maps of parts of some of the principal telluride mines of Boulder County, Colorado, showing structural controls of the ore.

as dams or baffles that impeded circulation of the ore solutions and caused deposition on the west side of the damming structure. On a local scale this is shown strikingly in the Ingram mine where the largest ore body was formed mostly under the gently dipping Fortune reef, though the solutions apparently leaked through locally (see Fig. 5B). On a somewhat larger scale, as shown in Figure 7, the Poorman center and Nil Desperandum center terminate against the Maxwell reef, and the Ingram center terminates against both the Bull o' the Woods and Maxwell reefs. On a still larger scale, it should be noted that in the southern part of the Boulder County telluride belt no ore was mined east of the Hoosier reef; in the central part (Gold Hill district) there was no ore east of the Maxwell reef; and in the northern part (Jamestown district) there was no ore east of the Standard reef. Thus it appears that the telluride solutions rose from depth along certain breccia reefs that overlay source areas, spread out into the vein fissures, and were dammed in places by other breccia reefs.

Some of the centers of telluride mineralization, as shown in Figure 7, also appear to have been centers of other types of mineralization in the belt, as most of the production of the lead-silver and pyritic gold ores has come from these localities. As noted earlier, some telluride ore has come from the

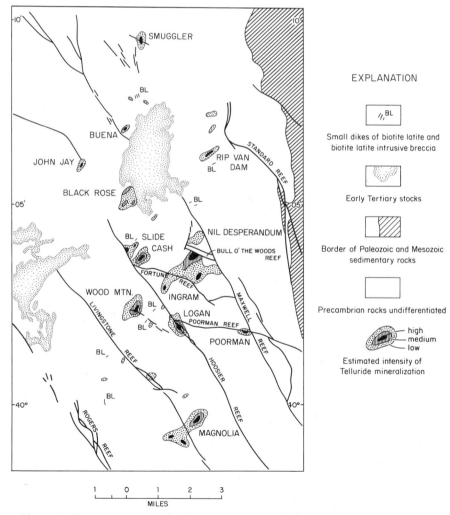

Figure 7. Structure map of the Boulder County telluride belt showing the relationship of productive centers to breccia reefs.

eastern part of the tungsten district and in the telluride belt, small amounts of tungsten ore were mined from the Logan and Magnolia centers.

In the Jamestown district, however, the fluorspar ores, and to some extent the pyritic gold and lead-silver ores, do not fit this pattern. Here, as pointed out by Goddard (1935, pp. 376–377), there is a rough zonal arrangement of the various types of ores around the Porphyry Mountain stock; the lead-silver ores are concentrated in a small area on the west side of the stock, the fluorspar ores overlap these but are a little farther out on the west and south sides, the pyritic gold ores are still farther out, and the telluride ores occur around the periphery of the district. It thus appears that the stock marked the center of the earlier ores and had an additional structural influence on the later telluride mineralization. It seems likely that available channelways in proximity to the stock were effectively sealed by the older lead-silver, fluo-rite, and pyritic gold deposits prior to introduction and circulation of the telluride fluids. Only in exceptional cases such as that of the Buena mine is there evidence that these fluids utilized the same fissures occupied by the older ores.

In this connection, it is of interest to note the control apparently exerted by the older and larger granodiorite stock which extends for several miles from its contact with the Porphyry Mountain stock in the north to the south-ernmost limits of the Jamestown district. No telluride deposits have been found within the limits of this stock, although it appears that some ores, par-ticularly those of the Black Rose center (Fig. 7), were localized by damming against this body. Compared to the Precambrian granites that most com-monly are the host rock of the ores, the granodiorite is a tough, well-knit, and finer-grained rock which behaved differently under stress and did not develop throughgoing fissures receptive to the ore fluids.

The source of the telluride solutions is not definitely known, but Lovering and Goddard (1950) and Lovering and Tweto (1953) make a strong case for a close genetic relationship to the biotite latite and biotite latite intrusion breccia. These dikes are small and inconspicuous but are widely scattered throughout the telluride belt (Fig. 7) and largely absent in the rest of the Front Range mineral belt. They are earlier than the telluride ores and younger than the pyritic gold ores. In the Logan mine, there is a striking spatial relation of high-grade telluride ore to upper parts of irregular dikes of intrusion breccia which were emplaced during the last stages of pre-telluride fault movement (Lovering, 1941). Lovering's diagram illustrating this relationship is reproduced as Figure 4. The biotite latite contains up to .01 ounce of gold per ton. Lovering (1941, p. 262) also mentions the unusu-ally high combined (4.06 percent) and uncombined (4.60 percent) water content of the fresh biotite latite as evidence of the volatile-rich and mineral-izing character of the original magma. Lovering and Tweto (1953) suggest that the tungsten ores of Boulder County are also genetically related to the biotite latites, but the evidence for this association does not seem to be as strong as for the telluride ores.

The general distribution of the biotite latite dikes as shown in Figure 7 seems to indicate the general extent of the source area of the telluride solutions, and the northerly trend and broad shape of the telluride belt seems to have been determined by the intersection of the northeastern end of the Front Range mineral belt with the relatively concentrated zone of breccia reefs. Quite probably fissures were formed elsewhere between the breccia reefs but were not within reach of the telluride solutions.

LOCAL CONTROL WITHIN VEIN FISSURES

The local structural control of ore bodies within the Front Range mineral belt have been discussed in detail by Lovering and Goddard (1950), but a few pertinent relations are noted below with reference to the telluride ores. The most significant factor is the presence of open ground along the fissure (either an open space or loose breccia filling) at the time the solutions were rising. Thus structural conditions that caused strong brecciation and fracturing were the most favorable. These include vein junctions, intersections of veins with earlier veins or faults, intersections of veins with igneous dikes, irregularity of the veins as related to wall movement, and the physical character of the wall rock.

The relative importance of these factors varied from one mine to another, and in some mines nearly all were effective. In the Buena mine, the "Big Stope" ore body was formed at the intersection of three veins at the contact of granite with a schist body (Fig. 6), and it was a common saying among the miners that wherever there was a junction of the Grant vein with an earlier vein, an ore body could be expected. In the Slide mine, the junction of the main vein with the Footwall vein seems to have been the controlling factor (Fig. 6), although the abrupt bend of these veins to a more easterly trend was significant. In the Smuggler mine, the intersection of the Smuggler vein with the earlier Careless Boy "vein" was the important structural feature (Fig. 6). In the John Jay, Emancipation, Ingram, and many other mines, ore bodies occurred at the junction or convergence of two branches of the same vein as illustrated by Figure 6.

In many of the veins, ore bodies were found in the more east-trending segments whereas the more northeast-trending portions were tight. On these vein fissures, as on most of those in the Front Range mineral belt, the southeast wall moved down and to the southwest at a moderate angle, thus producing more open ground in the more east-trending parts (Fig. 6).

Granite was by far the most favorable wall rock as it fractured and brecciated readily to form open ground, whereas the schist and hornblende gneiss were smeared out into tight, gougy shear zones. In a few places ore occurred in granite-gneiss, and in some ore bodies along granite-schist contacts ore extended short distances into the schist (e.g., Buena and Smuggler mines). Replacement played an insignificant role in ore localization and the chemical composition of the wall rocks was correspondingly unimportant.

The miners in the various telluride districts early recognized the control exercised by junctions, intersections, and abrupt changes in strike long before the geologists arrived, and the latter are deeply indebted to the local miners for their keen observations and helpful suggestions.

Identification of the Tellurides

GENERAL REMARKS

A variety of techniques was employed in identifying telluride minerals in the Boulder deposits, and the resulting data are summarized in this section. No attempt is made to repeat lengthy tabulations of X-ray data and microchemical reactions already available in the literature (Thompson, 1949; Berry and Thompson, 1962; Uytenbogaardt, 1951), but new data and tests found particularly useful for specific minerals are stressed.

X-RAY METHODS

The writers have relied most heavily upon X-ray powder methods for positive identification of the tellurides, and more than 200 samples have been so analyzed. Unknown minerals were extracted from polished sections with a dental drill and X-ray spindles prepared following procedures similar to those described by Hiemstra (1956). For much of the work, a Debye-Scherrer 57.3 mm camera was used, but a larger 114.6 camera was employed for samples requiring greater resolution. All diffraction work on the tellurides was performed with CuKα radiation (Ni-filter). Some of the X-ray data are given under specific minerals in the following section on Descriptive Mineralogy.

Qualitative X-ray fluorescence analyses of many mineral powders were obtained using a Norelco vacuum spectrograph, and in many instances these tests were performed on the same powders used for diffraction. A technique found generally useful was the insertion of entire polished sections in the Norelco unit. The polished sections were placed in a homemade aluminum sample holder that fits into the rotating chamber of the fluorescence unit. The sections were scanned for major components and detectable minor components, and these data were then correlated with microscopic information.

ELECTRON MICRO-PROBE ANALYSES

A limited amount of electron probe work was performed under contract by the Advanced Metals Research Corporation on an instrument in their laboratory in Somerville, Massachusetts. Included were quantitative analyses of minute phases which could not be identified by routine microscopic procedures and also traverses for selected elements across various telluride veinlets. Evidently difficulties were experienced when the tellurides, particularly hessite and to a lesser extent petzite, reacted to heat generated by the electron beam, and it was necessary to defocus the beam and work at low potentials unfamiliar to the AMR technicians. For these reasons, it is doubtful that the desired resolution was attained in some of the traverses and so such results will not be presented. However, analyses helpful in identifying unknowns will be cited in later pages.

OPTICAL ROTATION PROPERTIES

Data gathered on the rotation properties of selected anisotropic ore minerals of the Boulder veins are given in Table 2. The reader is referred to the works of Hallimond (1953), Cameron (1961), and Berek (1937) for the theory and methods dealing with the measurement of these properties.

All quantitative observations in the present study were performed in air and monochromatic light (586 mμ). The Hallimond method (Hallimond, 1953; Cameron, 1961) was employed in measurements of A$_r$, the apparent angle of rotation, and 2ϑ, the angular phase difference between components of the ellipse of vibration at 45° to the axes of that ellipse. Apparent angles of rotation were measured orthoscopically to the nearest 0.1 degree with a Nakamura plate inserted in the focal plane of a Wright slotted ocular that is fitted to a Leitz AMOP microscope. Ellipticities were determined conoscopically with 2ϑ values being computed from compensation settings of a rotary mica compensator placed in the tube slot of this microscope. The settings could be read to the nearest tenth of a degree. All values of A$_r$ and 2ϑ presented in Table 2 are based on numerous readings that were checked independently by several other observers.

The values of A$_r$ and 2ϑ listed in Table 2 are the values obtained directly from observation ("observed") and the values corrected for rotation of light

TABLE 2.   Rotation Properties at 586 mμ of Selected Tellurides and Other Ore Minerals of the Boulder County Veins

Mineral	Observed		Corrected		Effect of Gypsum Plate
	A$_r$	2ϑ	A$_r$	2ϑ	
Empressite	5.3 ± .2	+ 5.4 ± .4	4.4 ± .1	+ 4.5 ± .3	strong(+)
Ferberite	2.4 ± .3	+ 1.3 ± .4	2.0 ± .2	+ 1.1 ± .3	weak(+)
Hessite	0.90 ± .1	+ 4.6 ± .4	0.75 ± .1	+ 3.8 ± .3	indistinct
Melonite	1.0 ± .1	+ 2.9 ± .2	0.84 ± .1	+ 2.4 ± .2	weak(+)
Nagyagite	0.78 ± .1	+ 1.1 ± .4	0.65 ± .1	+ 0.9 ± .3	indistinct
Stuetzite	0.60 ± .1	+ 3.6 ± .3	0.50 ± .1	+ 3.6 ± .2	indistinct
Sylvanite	4.3 ± .1	− 3.5 ± .3	3.6 ± .1	− 2.9 ± .2	strong(−)
Tellurium	3.1 ± .2	− 1.6 ± .4	2.6 ± .2	− 1.3 ± .3	moderate(−)
Tetradymite	1.4 ± .1	+ 1.2 ± .3	1.2 ± .1	+ 1.0 ± .2	weak(+)

produced by upward passage through the reflecting place of the vertical illuminator ("corrected"). For our microscope the corrected values are 82.5 percent of the observed values.

Also listed in Table 2 under "Effect of Gypsum Plate" are qualitative observations of the sign and magnitude of the angular phase difference (Δ_{xy}) along the axes of the ellipse of vibration with the mineral in a 45° position. Procedures involved in this simple but useful test are described by Cameron (1961, p. 139).

The data presented in Table 2, which are based on work with X-rayed samples, agree well with the data previously published for these same minerals by Cameron and his co-workers (Cameron, 1961; Cameron and others, 1961; Carpenter and Cameron, 1963; Van Rensberg and Cameron, 1965). However, very inconsistent rotation data have been obtained from different X-rayed samples of krennerite and calaverite and these do not check with the previously published information. Additional measurements on specimens of known composition are needed and, for the present, these two minerals are omitted from Table 2.

POLARIZATION FIGURES

In a study of the rotation properties of certain tellurides prior to recent refinements of technique and equipment, Hase (1952) emphasized the diagnostic value of color effects that can be observed qualitatively in the polarization figures of these minerals. The authors have likewise found these effects a rapid and diagnostic means of distinguishing among some of the anisotropic tellurides. For examples, the dispersion fringes are strikingly different for stuetzite, hessite, and empressite, and the colors seen in the polarization figure of sylvanite immediately set this mineral apart from calaverite or krennerite. Some of the more distinctive figures observed are described in Table 3 which gives (1) the appearance of the figure with the polars precisely crossed and the mineral in a 45° position and (2) the appearance of the figure with the mineral in an extinction position and the analyzer rotated to bring the isogyres almost to the edge of the conoscopic field. All that need be done to observe these effects is to use a high power (unstrained) objective, cross the polars, and remove the ocular to examine the figure produced on the back surface of the objective. The dispersion effects will vary slightly from one microscope to another but the differences are small. The descriptions given in Table 3 apply equally well to figures produced by three different microscopes in our laboratory.

REFLECTIVITIES

Reflectivities of the tellurides and associated ore minerals were determined photoelectrically following procedures and using equipment recommended by Cameron (1961, 1963). A Photovolt Model 520M photometer, with photomultiplier tube No. 21C, was used in conjunction with a Leitz AMOP microscope for these measurements. The tube is attached to the top of a Leitz

TABLE 3.   POLARIZATION FIGURES OF SELECTED ANISOTROPIC TELLURIUM MINERALS
OF THE BOULDER COUNTY VEINS

Mineral	Appearance with Analyzer Precisely Crossed, Mineral at a 45 Degree Position	Appearance with Analyzer Slightly Uncrossed, Mineral at Extinction
Calaverite	Sharp black isogyres No distinct fringes	Isogyres dispersed No distinctive colors
Empressite	Indistinct gray isogyres Strong fringes, blue concave, yellow convex	Distinct gray isogyres Distinct blue concave White convex
Hessite	Distinct gray isogyres Strong fringes, rose-pink concave, green-blue convex	Distinct gray isogyres Indistinct blue concave White convex
Krennerite	Distinct gray isogyres No distinct fringes	Isogyres dispersed No distinctive colors
Melonite	Sharp black isogyres Strong fringes, yellow concave, white convex	Isogyres largely dispersed Pale blue concave Yellow convex
Stuetzite	Indistinct gray isogyres Orange-pink concave White convex	Indistinct isogyres Distinct blue concave White convex
Sylvanite	Fairly sharp black isogyres Blue-green concave, white convex Conspicuous pinks and yellow-greens in opposite quadrants when mineral close to extinction. Effect lacking in calaverite, krennerite and tetradymite.	Isogyres dispersed No distinctive colors
Tellurium	Distinct gray-black isogyres No distinct fringes	Isogyres dispersed No distinctive colors
Tetradymite	Distinct gray-black isogyres Pale green concave, white convex	Isogyres dispersed No distinctive colors

Makam camera which is firmly braced and mounted over the eyepiece of the microscope. An adapting plate was made to fit snugly in the back of the camera while providing a light-tight housing for the conical extension piece (window) of the photomultiplier tube. Clamped in place, the ⅜ inch window is normal to the optic axis of the microscope and lies in the back focal plane of the camera. A graticle with circular rulings was prepared and inserted in the side tube of the Makam camera to provide a means of centering the objective and to indicate the size and location of the mineral field actually projected to the sensitive surface in the photomultiplier tube. Most of the work was performed in air with a coated Leitz objective (45X, N.A. = .65) which allowed the testing of areas 80 microns in diameter without serious glare effects. In all measurements, the field iris was restricted and focused and the aperture diaphragm was closed to a fixed effective aperture of .08. The polarizer was set with its privileged direction vertical (i.e., parallel to the symmetry plane of the reflecting prism in the vertical illuminator).

Illumination was provided by a Leitz tungsten filament Monla lamp which was stabilized by a simple circuit in which the house current (110V, AC) was fed to a Sola constant voltage regulator (± 1 percent) whose output was further stabilized by a Sola step-down transformer placed in series between the regulator and the lamp. Two variable Ohmite resistors were placed in the circuit to provide a coarse and fine adjustment of current flowing through the lamp and the current was monitored by an ammeter which could be read to the nearest tenth of an amp and interpolated to .01 amps.

Reflectivity measurements were performed on newly polished and carefully leveled sections of the tellurides. For readings in "white light," the lamp current was set at a constant 5.1 amps for a color temperature of 2850°K. and all filters were removed from the optical system. A Schott narrow band interference filter peaked at 586 mμ with a half-value width of 13 mμ was used for measurements in light simulating sodium yellow. The linearity of the photomultiplier tube response was tested by inserting a rotating, adjustable, sectored disc between the vertical illuminator and the light source, and then recording the tube response as the light was reduced by known percentages by means of this light "chopper." No deviation from linearity could be detected at 586 mμ over the full range of light intensities to which this tube was subjected during the reflectivity measurements.

Bowie (1962) has discussed the shortcomings of photomultiplier tubes as applied to "white light" reflectivity determinations and his comments are applicable to the present instrument. While this instrument is highly sensitive and remarkably linear, there is no escaping the fact that the spectral sensitivity of the photomultiplier tube (peaked at 380 mμ) is substantially different from that of the human eye. For this reason, "white light" reflectivities measured relative to pyrite for dispersive minerals may be markedly different from those anticipated from visual observation or measured with a visual photometer. For example, surprisingly low "white light" reflectivities are recorded for native gold because this mineral is less reflective than pyrite in wave lengths to which the photomultiplier tube is most sensitive. While these effects do not detract from the reproducibility or quantitative value of "white light" reflectivities measured *with a given instrument,* they prevent meaningful correlation with visual photometry and can even cause discrepancies in results obtained with photomultiplier tubes of different spectral response. For these reasons, "white light" reflectivities of the tellurides are included in this report only to show the useful and reproducible spread of data helpful in identifying these minerals with a given photometer and different numbers might be expected with different instruments. On the other hand, reflectivity measurements in "monochromatic" light are not subject to these same errors provided that the light is truly monochromatic or covers a very narrow spectral band. It is for this reason that a narrow band interference filter was used for the measurements at 586 mμ and these results should be comparable to those obtained at the same wave length in other laboratories.

The writers have followed with great interest the deliberations of the I.M.A. Commission on Ore Microscopy with regard to various standards that might be applied in our work with the tellurides, but unfortunately could not await their final decisions. Even now, most of their proposed reflectivity standards (see Bowie, 1965, p. 1327) are provisional and must be calibrated in various wave lengths before application. Therefore, a polished diamond, mounted and leveled in dull black plaster, was chosen as a standard for work in monochromatic light. The choice was based on the fact that the true reflectivity of this surface (17.20 percent at 586 mμ) could be calculated from published values of n and k (Peter, 1923). This had the disadvantage of using a surface of low reflectivity for measurement of highly reflective minerals, but the exceptional linearity of the photomultiplier tube offset this problem. For "white light" reflectivities, a natural face on Elba pyrite was used and assumed to have a reflectivity of 54.6 percent based on measurements by Bowie (1962).

Reflectivity data for the more unusual minerals of the Boulder veins are presented in Table 4. Certain minerals could not be included (e.g., nagyagite, rickardite) because the available grains were either too small or too impure for measurement. The mean reflectivities given are the numerical averages of extreme readings obtained on at least 30 mineral grains for each species. For the isotropic minerals, this range reflects compositional variations and

TABLE 4.   REFLECTIVITIES IN "WHITE LIGHT" AND AT 586 mμ OF SELECTED
TELLURIDES AND OTHER ORE MINERALS OF THE BOULDER COUNTY VEINS

Mineral	"White Light" 2850°K *			586 mμ †		
	Mean	Range	Bireflec-tance**	Mean	Range	Bireflec-tance**
Altaite	62.3	60.0–64.6	—	62.8	60.0–66.2	—
Calaverite	55.3	52.1–59.5	9.1	63.6	56.2–66.2	10.0
Coloradoite	37.6	37.0–38.5	—	38.6	37.6–39.3	—
Empressite	42.5	35.0–50.0	25.8	47.0	37.0–55.5	28.8
Ferberite	16.7	15.6–18.0	9.4	18.1	16.0–18.9	9.0
Native Gold (>900 fine)	65.5	63.1–69.0	—	85.5	83.5–86.9	—
Hessite	40.6	39.0–42.5	3.1	37.8	36.3–40.3	3.2
Krennerite	54.5	49.9–58.5	5.8	56.6	51.5–61.5	8.9
Melonite	54.6	53.5–55.9	1.8	59.8	59.0–61.1	2.9
Petzite	40.6	38.9–43.2	—	38.8	36.0–39.5	—
Stuetzite	40.9	39.9–42.0	2.9	38.2	36.8–39.0	2.1
Sylvanite	50.2	44.0–57.0	18.3	54.9	47.0–61.8	19.4
Tellurium	59.8	56.0–64.2	10.4	62.7	59.9–67.9	11.6
Tetradymite	51.0	48.5–52.9	4.0	54.2	51.3–57.0	7.3

*Based on pyrite standard assumed to have a reflectivity of 54.6% at 2850°K.

†Based on polished diamond standard having a calculated reflectivity of 17.20% at 586 mμ.

**The maximum variation in reflectivity of a single grain expressed as a percentage of its maximum reflectivity.

slight imperfections of polish, whereas the range indicated for anisotropic minerals results from bireflectance in addition to these other variations. The greatest variation in reflectivity displayed by a single grain of the mineral upon rotation of the stage is given under "bireflectance" and is expressed as a percentage of the maximum reflectivity of that grain. The bireflectances are valuable in distinguishing among the tellurides, many of which have fairly close mean reflectivities. For example, the mineral empressite (AgTe) has enormous bireflectance which instantly distinguishes it from all other tellurium minerals in the list. Likewise, sylvanite is conspicuously more bireflectant than either calaverite or krennerite and, at least in the Boulder ores, there is no problem of distinguishing the sylvanite from these other minerals.

Some of the tellurides are very prone to tarnish and most are easily scratched, so it is essential that the reflectivities be determined on freshly polished surfaces. Altaite presents an extreme example. This mineral is bright white when newly polished (white light reflectivity = 62.3 percent) but tarnishes gradually in air to a yellow white (54 percent), yellow (40 percent), and, in extreme cases, a dull gray (35 percent). Only complete repolishing can restore the original color and reflectivity.

Another source of variation in reflectivity is the state of aggregation of the unknown mineral. For example, melonite typically occurs in very fine-grained aggregates that polish to a variably pitted surface. The pitting, of course, lowers the reflectivity and at the same time tends to introduce false bireflectance. In unpitted areas, the numerous grains of diverse orientation give minimum values for bireflectance. For these reasons, the data given for melonite in Table 4 are only those obtained for comparatively rare large single crystals of the mineral, rather than the typical fine-grained aggregates. Identical problems arise in reflectivity measurements of fine-grained, porous aggregates described as "sponge tellurium" in later pages. Again, the data for tellurium in Table 4 were obtained on large single crystals of the mineral.

INDENTATION MICROHARDNESSES

Vickers indentation hardnesses of the tellurides and other ore minerals were determined with the GKN Hardness Tester, an instrument manufactured by Associated Automation, Ltd. The characteristics and operation of this instrument are described by Bowie and Taylor (1958), who applied it in a survey of numerous metallic minerals. Our instrument is fitted to a Cooke, Troughton, and Simms Metalore microscope that rests on a sturdy concrete pier.

The test load found most satisfactory for general work with the Boulder tellurides was 30 grams. Lighter loads produced marks too small to be measured with precision and heavier loads produced indentations so large that suitable test areas could not often be found in the fine-grained telluride intergrowths. For some of the minerals like nagyagite, which occur rarely and as very tiny grains, even the 30 gram weight was too heavy and, of necessity, tests were performed with a 3 gram weight.

Published indentation hardnesses for the tellurides are rather sparse and so the data obtained for these and other rare minerals in the Boulder veins are given in Table 5. For each mineral, the test load, mean Vickers Hard-

TABLE 5.   VICKERS INDENTATION HARDNESSES OF SELECTED TELLURIDES AND OTHER ORE MINERALS OF THE BOULDER COUNTY VEINS

Mineral	Test Load (grams)	Number Diagonals Measured	Mean H_V	Range H_V	Character of Indentation
Altaite	30	90	32.0	19.8–53.3	Excellent, undistorted mark. No fracturing.
Calaverite	30	34	215	153–406	Slight distortion. Irregular fractures radiate from some marks.
Coloradoite	30	105	30.7	20.4–51.1	Excellent, undistorted mark. No fracturing.
Empressite	30	26	151	95.8–204	Good mark. Slight plastic mounding. Few cleavage (?) cracks.
Ferberite	30	28	374	226–535	Some heaving along radial and concentric cracks. Cleavage not evident.
Gold	30	34	42.0	23.5–58.3	Good mark. Concave outward due to slight plastic recovery.
Hessite	30	40	17.4	11.4–27.7	Excellent mark. Slight curvature. No fractures.
Krennerite	30	24	136	75.2–315	Moderate distortion. Fracturing along one cleavage direction as distinct from calaverite.
Melonite (fine-grained)	30	24	63.1	47.8–91.9	Slight distortion. Cracks follow grain contacts.
Melonite (coarse crystals)	30	24	49.3	33.7–78.6	Moderate to strong distortion. Slippage along basal cleavage.
Nagyagite	3	14	30.1	10.4–63.0	Moderate to extreme distortion. Opening and offsets along 010 cleavage direction.
Petzite	30	72	46.7	35.1–52.7	Excellent, undistorted mark. No fracturing.
Stuetzite	30	28	92.7	45.4–117	Good mark. A few, gently curving fractures spread from indentation.
Sylvanite	30	43	89.7	35.1–151	Variable distortion. Slippage along cleavage direction.
Tellurium (sponge variety)	30	12	21.1	12.1–36.6	Good mark. Some irregular fracturing.
Tellurium (coarse crystals)	30	74	54.0	20.8–86.9	Slight to moderate distortion. Offsets along cleavage. Edges slightly concave outward.
Tetradymite	30	38	32.9	17.4–73.6	Slight to extreme distortion. Shattering and offsets along cleavage direction.

ness, and range of Vickers Hardness are indicated along with a brief description of the character of the mark typically produced. The latter is especially important because the appearance of the indentation is commonly more diagnostic than its size. This is the case, for example, with calaverite and krennerite whose microhardnesses overlap. Calaverite lacks cleavage and tends to develop irregular fractures that radiate outward from the indentation, whereas krennerite often develops a set of close-spaced parallel cleavage fractures along the edges of the indentation.

Even if one rejects tests where the size and shapes of the indentation are grossly distorted by fracturing, heaving, or extreme plastic distortions, indentation hardnesses may vary over wide ranges for a given mineral. This is illustrated in Figure 8 which shows the frequency distribution of indentation hardnesses obtained for minerals in the Boulder suite. There is considerable overlap of hardnesses for these minerals and the dangers inherent in basing identifications on the size of just a few indentations are immediately apparent. Wherever possible a series of indentations were made in both the unknown and an adjacent known mineral for relative values and these data were then used in conjunction with other properties rather than as a primary basis for identification. Also noteworthy in Table 5 and Figure 8 is the effect of grain size and state of aggregation on microhardness. The fine-grained aggregates of melonite in the Boulder ores are "harder" than single crystals of the same mineral whereas fine-grained aggregates of tellurium are "softer" than large single crystals. In the case of melonite, aggregates of small interlocking grains are stronger than a large single crystal. In the case of tellurium, the fine aggregates are quite porous and the diamond indentor emplaces itself to some degree by collapsing these pores, which evidently requires less stress than indentation of a large, solid grain of the same mineral.

MICROCHEMICAL TESTS

After applying standard microchemical etch tests (Short, 1940) to the tellurides identified by X-ray methods, a few dependable and diagnostic tests were recognized and subsequently applied to many samples. These few tests are described in the following paragraphs under "Diagnostic Properties of Individual Tellurides." Much more extensive use of etch reagents was applied to structural etching and to development of visual contrast between similar minerals to facilitate modal estimates and photography of the tellurides. Reactions useful for these purposes are also described in the following pages.

Chemical contact printing (Galopin, 1947; Gutzeit, 1942; Williams and Nakhla, 1951) was very well suited to certain special problems encountered with the tellurides. For example, this method offered a rapid means of spotting the presence and distribution of melonite in the ores prior to selection of materials for polishing. Direct prints for nickel were taken from rough sawed slabs of ore using nitric acid as the attacking solvent, ammoniating the paper by holding it in fumes of concentrated NH_4OH, and

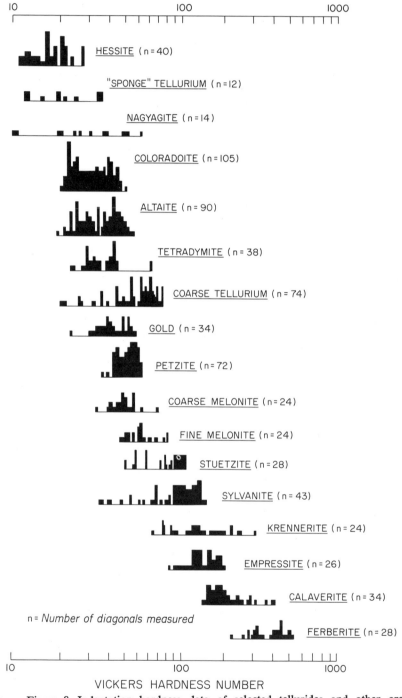

Figure 8. Indentation hardness data of selected tellurides and other ore minerals.

developing the latent image with dimethyl glyoxime (2 percent solution in alcohol). The location of any melonite in the sample is revealed by positive red areas appearing in the print. Similar tests were used to determine the distribution of other elements including copper, lead, bismuth, silver, and sulfur and were quite helpful in preliminary selection of chips for polishing.

A number of other spot tests employed in this study will be mentioned briefly in later pages in connection with the particular minerals or specific problems to which they were applied.

DIAGNOSTIC PROPERTIES OF COMMON TELLURIUM MINERALS

General Remarks

With few exceptions, each of the tellurium minerals present in the Boulder ores has certain easily observed properties that serve to distinguish it from all others. These diagnostic properties are described on the following pages with emphasis on those found helpful in distinguishing between the most similar and difficult pairs. Wherever possible, simple yet dependable tests that require the least amount of time, experience, and equipment are stressed. No attempt is made to reproduce systematic lists of colors, etch reactions, and other properties, which are already published (Thompson, 1949; Galbraith, 1940; Uytenbogaardt, 1951); many of these properties are of little or no immediate help in the identifications.

Altaite

There are many soft, white or creamy-white tellurium minerals, including calaverite, krennerite, sylvanite, tellurium, and tetradymite, that could possibly be mistaken for altaite were it not for the fact that all of these are distinctly anisotropic. Altaite is the only very white, highly reflective, *isotropic* mineral in the Boulder telluride ores. Compared to altaite, the other soft, isotropic minerals, including galena, coloradoite, and petzite, have much lower reflectivities and appear gray or gray-white. Tetrahedrite is darker, duller, and much harder than altaite. Occasional triangular cleavage pits may appear in altaite, just as they do in petzite, but these are not nearly so common as they are in galena.

Altaite is also unique in its tendency to tarnish more rapidly than other tellurium minerals on prolonged exposure to air or brief exposure to acid fumes. After several weeks of exposure, altaite gradually turns to a permanent yellow or gray (Pl. 7f) and the tarnish commonly brings out a mosaic-granular texture not seen in other tellurides. The effect can be accelerated by exposing the section to fumes of concentrated nitric acid. A practical way to develop strong contrast between altaite and native tellurium or other white tellurides is quickly to dip the polished section in 50 percent HCl or 1:7 HNO_3 and immediately wash it. Altaite is etched a flat gray while the other minerals are unaffected (Pl. 5a).

Calaverite-Krennerite-Sylvanite

As a group, these minerals share certain properties that quickly distinguish them from other ore minerals, but they are very similar to one another. All are white to creamy-white, soft, highly reflective, anisotropic, and, most important, react distinctively to nitric acid by turning a tan-bronze color and developing one of several etch cleavage patterns described in detail by Short (1937). This test is a very reliable criterion of the group, but the writers have had little success in using it to separate the individual minerals.

Sylvanite can be easily distinguished from other members of this group on the basis of its much stronger bireflectance (Table 4) and very distinctive polarization figure (Table 3). Many authors (Ramdohr, 1950; Thompson, 1949; Sindeeva, 1964), including the present ones, believe that the strong multiple twinning seen very commonly in sylvanite sets that mineral apart from either calaverite or krennerite (Pls. 3e and 9b). However, the dependability of this particular property is open to question because Cabri (1965b) has now found such twinning in one grain of X-rayed krennerite. A few of nine X-rayed krennerites from Boulder County display occasional twin boundaries, but not polylamellar twinning. No twinning of any kind was observed in 28 calaverites that were X-rayed.

Except for sylvanite, microscopic identifications of other members of this group are at present difficult and questionable. The problem is compounded by the existence of still another member of this group, montbrayite (Au_2Te_3), a mineral very similar to krennerite (Peacock and Thompson, 1946). While montbrayite is known only in one locality and does not occur in the Boulder veins, there is no *a priori* assurance that it is not present in other localities. About all that can at present be done with the microscope is to look for cleavage in the remaining minerals of this group. If cleavage is present or can be induced by indentation, then the mineral is either krennerite or montbrayite but cannot be calaverite. Optical rotation properties offer future promise of reliable criteria, but tests must first be performed on calaverites and krennerites of varied and known composition and montbrayite must also be studied. Sindeeva (1964) relies on microhardness, pointing out that calaverite is much harder than either krennerite or sylvanite. Reference to Table 5 and Figure 8 will show this is in general true, but also that there is extensive overlap in the microhardness ranges for these minerals. Furthermore, the indentation hardness of montbrayite is unknown.

Kostovite, a recently discovered gold-copper telluride ($AuCuTe_4$), is another mineral optically similar to sylvanite, but it is negative to all standard etch reagents (Terziev, 1966). This mineral has not been found in the Boulder ores.

In view of the uncertainties with this group of similar minerals, we have routinely X-rayed *all* samples of possible calaverite, krennerite, montbrayite, and untwinned sylvanite observed in the Boulder ores.

Coloradoite

Coloradoite resembles several other minerals in the ores but can be positively identified without recourse to X-ray methods. It is most similar to petzite and galena but its reaction to 1:7 nitric acid is entirely and consistently different. When this acid is applied, coloradoite develops a barely perceptible tan tint, but is essentially negative and retains its original polish and reflectivity for *at least one minute*. After that time, sometimes several minutes later, coloradoite begins a slow but distinctive reaction in which a permanent iridescent pitting begins at the edges of coloradoite areas and slowly moves inward. No etch cleavage develops. In contrast, galena is etched black within a few seconds and petzite turns a dark gun-metal gray within 60 seconds. The petzite reaction is further characterized by the formation of an unusual system of etch cleavages that look like mud cracks and appear after the section is washed and air-dried. This reaction is delayed where petzite is in contact with sylvanite, calaverite, or krennerite, but still takes place after vigorous reaction of those minerals is complete. The contrast that can be developed between coloradoite and petzite by applying nitric acid is illustrated by Plate 6b. The only other isotropic mineral with which coloradoite might be confused is tetrahedrite which is a much harder mineral of higher relief. A casual observer might mistake hessite or stuetzite for coloradoite but, among other differences, these minerals are distinctly and colorfully anisotropic.

Empressite

The name empressite is applied here to AgTe as redefined by Honea (1964). This mineral possesses extraordinary optical properties that are not shared by any other telluride. Its bireflectance and reflection pleochroism (cream-white to dark gray with a violet tint) are exceptionally large. The anisotropy is accordingly very strong with conspicuous blue, pinkish-brown, green, and white polarization colors. Measurements in monochromatic light do not do justice to the great dispersive properties of this mineral, but the phase difference and apparent angle of rotation measured at 586 mμ are greater than for any other tellurium mineral included in the present measurements. Etch tests and X-ray analyses are superfluous in the identification of this mineral, although we have confirmed it by X-ray methods in the Boulder veins. The spectacular optical properties are not likely to be confused with those of any other ore mineral.

Hessite

Hessite is anisotropic with definite reddish-brown and blue-gray polarization colors and a very distinctive polarization figure (Table 3). There have been occasional reports of isotropic hessite or of hessite with "anomalous anisotropism" that are misleading and should be evaluated in view of the importance of this property in identification.

At room temperature, hessite is possibly orthorhombic (Rowland and Berry, 1951), probably monoclinic (Frueh, 1959b), but certainly anisotropic.

Therefore, as pointed out by Cabri (1965a), descriptions of its anisotropism as "anomalous" (e.g., Stillwell, 1931; Edwards, 1954; Baker, 1958; Callow and Worley, 1965) are incorrect because the mineral is indeed anisotropic. At temperatures above 145°C (in the binary system Ag-Te; Kracek and others, 1966), hessite is cubic (Frueh, 1959b) but inverts to the ortho- rhombic or monoclinic form upon cooling. In the process, it may retain cubic crystal *forms* but not the cubic *structure* which is unquenchable (Frueh, 1959b; Cabri, 1965b, Kracek and others, 1966). Uytenbogaardt (1951, p. 53) attributes the spotted appearance of extinction in some hessite to struc- tural relicts of the cubic hessite, but this seems untenable in the light of quenching experiments. We regard the spotted extinction common in hessite as due to an interpenetration of irregular twins of diverse orientation. We have seen twinning of this kind (Pl. 6f) in which a series of hessite grains may remain at extinction or uniformly illuminated as the stage is rotated be- cause of the orientation of the single twin they represent. Baker (1958, p. 25) describes hessite from the Phantom Lode at Kalgoorlie that shows "anoma- lous anisotropism" and notes that some of the grains are isotropic while others display a patchy extinction. He does not mention specifically X-raying the isotropic material. Further tests might prove this to be petzite, a mineral commonly intergrown with hessite, but which Baker did not report in these ores.[3] To our knowledge, there has never been a case of isotropic hessite (natural) that has been firmly established by X-ray methods. On the other hand, there are many cases of X-rayed hessites that are structurally and optically anisotropic. For these reasons, the authors regard the anisotropy as a reliable property useful in the identification of hessite. It should perhaps be stressed that the anisotropy could be missed altogether unless sought in strong illumination. By removing the lamp filter and inserting the prism reflector, the effect is obvious at any magnification in air or oil and is spec- tacular if the analyzer is turned a degree or two from the crossed position (see Pls. 6e and 6f).

Hessite typically displays one of several twinning patterns that *may* be produced by inversion to the low temperature form (Pls. 6e and 6f). The interpenetrating cauliflower-shaped twins that produce a spotted or patchy extinction were already mentioned. The twins also occur in from one to four sets of parallel lamellae which may be straight or lensatic and measure up to .02 x .3 mm. Where two or more sets are present, they form a complex gridwork (Pl. 6e).

While there are other more subtle characteristics of color and polish, the anisotropy of hessite immediately distinguishes it from other soft tellurides like petzite and coloradoite that are isotropic. Also, hessite reacts rapidly to dilute nitric acid, turning dark iridescent or black, a reaction different from

---

[3]In a note published after submission of the present manuscript, Cabri (1967) re- ported the occurrence of petzite in one sample from the Phantom Lode, and confirmed its identity by X-ray methods.

those described under "petzite" and "coloradoite." Compared to other anisotropic minerals like sylvanite, tetradymite, tellurium, and so forth, hessite is much darker and shows a very different polarization figure (Table 3). Inclusions of nagyagite in hessite might be overlooked in ordinary light, but several distinct differences between these minerals are described under "nagyagite."

The mineral stuetzite (Ag_5Te_3; Honea, 1964) is very similar to hessite and care must be taken to distinguish the two in polished section. The similarity is especially close where hessite lacks its usual lamellar or patchy twinning. Fortunately, the polarization figures of these minerals are very different (Table 3) and offer a rapid and reliable means of identification. Other dependable differences are described under "stuetzite."

Melonite

Melonite does not closely resemble any other mineral in the Boulder ores. It has an unusual pink-white color in ordinary light and under crossed polars displays unique polarization colors with attractive shades of yellow and purple. The polarization figure (Table 3) is also diagnostic. Aggregates of fine-grained melonite (Pls. 7h and 8b) are resistant to polishing and stand in high relief against the other tellurium minerals, while coarse crystals polish well and display less conspicuous relief (Pl. 8f). The contact print test found helpful in locating and identifying melonite was previously described. Frohbergite, the iron telluride (Thompson, 1947), bears some similarity to melonite but displays orange-red to inky-blue polarization colors and, in contrast to melonite which stains dark-brown, frohbergite is inert to $FeCl_3$ (Uytenbogaardt, 1951). Frohbergite was sought in the Boulder ores but evidently is not present.

Nagyagite

In the Boulder deposits, nagyagite occurs consistently as small platey crystals intergrown with and normally enclosed by either petzite or hessite (Pls. 8h and 9a). In ordinary light, the color and reflectivity of nagyagite is sufficiently close to those of the host minerals that the minute inclusions could easily be overlooked. However, their presence is brought out by application of dilute nitric acid. The nagyagite is more resistant to etching and remains negative as petzite and hessite darken (Pl. 9a). Under crossed polars, nagyagite is distinctly anisotropic in contrast to petzite which is isotropic. The polarization colors of nagyagite are rather dull grays with pale tones of green and blue in contrast to the more lively red and blue shades associated with hessite or stuetzite. The indentation microhardness of nagyagite is extremely low (Table 5) and the mineral characteristically fails by splitting and slippage along one strong cleavage direction. The distortion is extreme and measurable indentations are difficult to obtain.

Petzite

Literature dealing with the optical properties of petzite, particularly its isotropism, is inconsistent and must be evaluated before turning directly to problems of identification. Helke (1934) described two varieties of petzite that were intergrown in ores from several districts including Boulder County, Colorado (no specific localities given). One of these he described as an isotropic, gray-violet mineral (alpha-petzite) while the other was reportedly very similar to hessite but a shade darker and characterized by a weak reflection pleochroism detectable in air (beta-petzite). He suggested a possible paramorphic relationship between these forms with the isotropic variety metastably preserved at room temperature by a sluggish transformation. Borchert (1935) attempted to find Helke's beta-petzite in ores from the same localities (including Boulder County), but was unable to do so. He did find anisotropic hessite intimately intergrown with petzite, but no other anisotrophic phase bearing any resemblance to petzite. Although he could not confirm Helke's findings, Borchert did not discount the possibility of the beta-phase nor that it might bear a paramorphic relation to isotropic petzite. Since these early works, there have been numerous descriptions of petzite as isotropic and some are backed up by X-ray identifications (e.g., Thompson, 1949; Cabri, 1965b; Markham, 1960; Frueh, 1959a). On the other hand, anisotropic petzite has never been established by X-ray methods, and only one new questionable occurrence has been reported since Helke's early work. Hawley (1948, p. 111) found petzite in one of 60 samples from Kirkland Lake, and reported that certain grains of the petzite were distinctly anisotropic and showed a fine grid of twin lamellae at about right angles. Hawley used X-ray methods to confirm the minor minerals in these ores, but made no claim to having X-rayed these minor grains of anisotropic petzite. From his description, particularly of the twinning (see Pl. 6e), the writers suggest that this material is hessite, a mineral which Hawley did not otherwise record in the ores.

Over the years, there have been many compilations of the properties of the ore minerals in which it is sometimes difficult to tell whether the information listed has been verified by the author or is merely a quotation from older sources. For example, Sindeeva (1964, p. 104) states that petzite is "partially anisotropic," but gives neither an explanation nor a basis for this statement. Ramdohr (1950) and Uytenbogaardt (1951) make it clear that their descriptions allow for Helke's anisotropic petzite, and Schouten (1962) in turn refers to Uytenbogaardt and Ramdohr as sources. It appears, therefore, that the notion of anisotropic petzite is firmly emplanted in the literature where it took seed from the single publication by Helke (1934) and has been nourished ever since by repeated publication, but not by repeated observation or X-ray confirmation.

Comparatively recent studies have not conclusively resolved this problem. X-ray analyses of natural petzites invariably reveal a cubic structure (Thompson, 1949; Frueh, 1959a; Cabri, 1965b). A polymorph of petzite of unknown structure has been found stable between 210° and 319°C (Cabri, 1965b; Frueh, 1959a), thus lending support to Helke's speculations at least to the extent that thermal modifications of petzite structure do take place. Cabri

(1965b) was able to quench the intermediate temperature form for X-ray analysis but does not describe its appearance in polished section. Evidently, Cabri did not find this metastable phase in natural ores which he examined for comparison with the synthetic products. Markham (1960), likewise, did not find this phase in X-ray studies of the Kalgoorlie and Vatukoula ores and did not record it even in synthetic runs that were apparently cooled more slowly than the quenched charges of Cabri (1965b).

In the present work, all polished sections were closely examined with the possibility of some anisotropic variety of petzite in mind. Isotropic petzite was observed in 70 polished sections and confirmed as cubic petzite by X-ray analysis of seven samples. All 70 samples were etched in several places with dilute nitric acid in the search for some inconspicuous phase that might be intergrown with the isotropic petzites but none was found. An additional 18 sections containing hessite (without normal petzite) were also studied in this way but again with negative results. The only anisotropic minerals bearing any resemblance to hessite were identified as stuetzite, pyrargyrite, and nagyagite. These phases were all confirmed by X-ray analysis and have distinctive microscopic properties of their own. Pyrargyrite is immediately apparent from its red internal reflections, and the properties of stuetzite and nagyagite are discussed elsewhere in this report.

The present investigation leads to the position taken 30 years ago by Borchert (1935), although more evidence has been gathered to support his stand. No anisotropic petzite has been found in the Boulder ores, but its possible existence cannot be dogmatically ruled out in other deposits. The petzite that does occur in the Boulder ores is consistently isotropic and can be readily distinguished from other gray-white minerals like hessite or stuetzite which are consistently anisotropic. The polarization figure of petzite is that of an isotropic mineral and is strikingly different from the anisotropic figures of hessite and stuetzite (Table 3). Other differences of color and polish are more subtle and can be used confidently only after some practice with a set of known minerals established by X-ray methods.

Petzite closely resembles coloradoite in that both minerals are isotropic, gray-white, and take a very good polish. However, the nitric acid test described under coloradoite is a very dependable one regardless of the variety of metallic minerals with which petzite and coloradoite are intergrown. Visual contrast between coloradoite and petzite can also be produced with a saturated solution of $HgCl_2$ which slowly turns petzite brown but does not affect coloradoite.

Galena might at first be mistaken for petzite, but it is somewhat whiter and displays much more pronounced cleavage. Galena is definitely distinguished from petzite by the fact that it is inert to $HgCl_2$ while petzite turns brown. Its reaction to 1:7 nitric acid (rapid blackening) is entirely different from the reactions already described for both coloradoite and petzite.

Sindeeva's comments (1964, p. 104) on the microscopic properties of petzite are misleading. They are in part as follows:

In its optical properties, hardness and response to standard reagents, petzite is not easily distinguishable from argentite, empressite, coloradoite, and tetrahedrite. It is similar to hessite to such an extent as to preclude a reliable distinction.

Petzite is as dissimilar to these varied minerals as they are to one another. Argentite, "empressite" (whether AgTe or Ag_5Te_3), and hessite are all anisotropic minerals with diagnostic polarization properties. None of these assorted minerals gives the characteristic petzite reaction to HNO_3. Hardness alone is enough to distinguish petzite from tetrahedrite whether tested by scratching or by indenting the minerals. Tetrahedrite in the Boulder ores ranges in Vickers Hardness from 202 to 452 (36 indentations, 30 gram load) in comparison with a range of 35.1 to 52.7 for petzite (36 indentations, 30 gram load).

Stuetzite

The name stuetzite is here applied to Ag_5Te_3 as recently redefined by Honea (1964). The only problem likely to be experienced in the identification of stuetzite is its great similarity to hessite, especially where hessite lacks its usual complex twinning. Both of these minerals are soft and have similar gray-white colors, similar polarization colors, and about the same degree of anisotropy. However, the polarization figures are very different (Table 3) and offer a means of immediate identification. Close inspection of the two minerals also shows other more subtle but dependable differences. Stuetzite takes a better polish than hessite and possesses greater abrasion hardness (evidenced by scratching), greater polishing hardness (evidenced by relief against hessite), and greater indentation hardness (Table 5). Stuetzite is a brittle mineral while hessite is sectile. This difference was revealed in attempting to drill out small samples of each mineral for X-ray analysis. Though similar, the colors of these minerals in reflected light are not exactly the same. In air, hessite is gray-white with a distinct tannish tint and pleochroism is barely perceptible while stuetzite is weakly pleochroic with colors ranging from blue-white (similar to a pale pyrargyrite) to a color indistinguishable from hessite. In oil, the differences are somewhat more pronounced but still weak, with hessite showing a change from tannish-gray to gray-white and stuetzite a stronger variation from dull blue-white to dark tannish-gray (similar to coloradoite). There are also very slight differences in the polarization colors of these minerals, but these are too small and subjective to be of help. Where present *in contact,* hessite and stuetzite react at different rates to dilute HNO_3. If the reagent covers both minerals, the acid first darkens hessite (Pls. 7b and 9f) and then proceeds to react more slowly on stuetzite ultimately turning both minerals black.

The polarization figures are recommended as the most conspicuous and dependable criteria of these two minerals.

Tellurium

Tellurium is included here because of its similarity to, and common associations with, the telluride minerals. This mineral is soft, white, highly reflective, and could be mistaken for several of the anisotropic tellurides with which it is commonly intergrown. It is easily distinguished from the calaverite-krennerite-sylvanite group by the nitric acid test. Tellurium effervesces vigorously and turns black without etch cleavage while the gold tellurides undergo their typical reaction as previously described. Striking contrast between tellurium and these tellurides can be obtained by dipping the section very quickly in concentrated nitric acid and immediately washing the surface. The results are illustrated in Plate 10e where tellurium is rendered a dark iridescent blue-black, sylvanite a bronze color, and coloradoite is bright and unetched. Altaite is isotropic and should not be confused with tellurium. Tetradymite is similar, but against tellurium it is more yellow and is always less reflective and more weakly anisotropic. Etch tests useful in distinguishing tellurium from altaite (see Pl. 5a) and from tetradymite (see Pl. 3b) are discussed under the headings for those minerals.

Tetradymite

A soft, white anisotropic mineral differing from native tellurium and the common tellurides was detected in six of the 195 polished sections investigated. From its etch reactions (strong $FeCl_3$ and HNO_3 reactions as described by Thompson, 1949) and general optical properties, the writers suspected that the unknown was one of the bismuth tellurides but could not tell which one. It was therefore necessary to X-ray all of these occurrences, and this showed the unknown to be tetradymite in every case.

While criteria can be offered for the distinction of tetradymite from other minerals that occur in the Boulder ores, these criteria do not pin down the identity of tetradymite as distinct from a host of very similar bismuth tellurides with which the authors have had no firsthand experience.

In the Boulder ores, tetradymite occurs as elongate platey crystals with tattered outlines solidly embedded in hessite and petzite (Pl. 11b), and in places these crystals are bent, revealing one very strong cleavage. Tetradymite has a flat white color against the darker gray-white of petzite and hessite. In other ores, the tetradymite is massive granular and the cleavage is not conspicuous. The mineral is very soft and it is difficult to obtain measurable indentations without heaving and slippage along the cleavage direction. In this respect, tetradymite resembles nagyagite, but the latter is a much darker gray.

Tetradymite is moderately bireflectant in yellow light, but its reflection pleochroism in "white light" is barely perceptible. The color is white when compared to duller, gray-white minerals like galena, petzite, hessite, and coloradoite, but a distinct yellow-white when seen against the more reflective white minerals like tellurium. Tetradymite is distinctly but not strongly anisotropic, and shows rather unimpressive white, gray, and black shades under crossed polars. It is immediately distinguished from altaite by its anisotropy and lower reflectivity, and from tellurium which is more strongly anisotropic and possesses much higher reflectivity. The simple test with the gypsum plate (Table 2) reveals a weak positive phase difference for tetradymite in contrast to a moderate negative phase difference for tellurium.

Certain etch tests are also diagnostic. Dilute nitric acid rapidly turns tetradymite iridescent and then black—a test that suffices to distinguish this mineral from somewhat similar krennerite and calaverite whose reactions have been described. The etch reactions of tellurium and tetradymite are similar, and so to develop visual contrast between these minerals the writers normally cover both with a very large drop of water to which a very small amount of dilute nitric acid is added. By increasing the acidity gradually, the point is reached where tetradymite begins to turn yellow while tellurium remains negative (see Pl. 3b).

Descriptive Mineralogy

GENERAL REMARKS

The minerals identified in the telluride veins are grouped in Table 6 according to their abundance and chemical composition. Sixty-seven species are involved either as principal metallic minerals (16), principal gangue minerals (11), or as comparatively minor or rare minerals of spotty occurrence (40). The mineralogical complexity of these ores is illustrated by the fact that any random sample is likely to contain a dozen or more of these vein minerals in fine-grained and intimate intergrowth.

This is the first such compilation for the Boulder County deposits based primarily upon X-ray identifications, and includes a number of minerals not previously recognized in these ores. As indicated in Table 6, all species listed were X-rayed with the exception of those whose identities were obvious, and a few others which occur in minute quantities. Some 220 samples were X-rayed in the process of compiling Tables 6 and 7.

Listed separately in Table 6 are the common wall rock alteration products described in other studies (Lovering and Goddard, 1950; Bonorino, 1959) and a number of metallic minerals formed during the lead-silver stage of mineralization which are associated with the tellurides in some ores. Some of the minor vein minerals listed have been reported in the early literature, but were unconfirmed in the present study.

Approximate mineral abundances and mine locations for each of the 156 tellurium-bearing polished sections studied are listed in Table 7. The abundances listed are based upon visual inspection of the samples and not point counts. The percentages listed for each polished section in Table 7 are the estimated volumetric percentages of each mineral making up a part of the total metallic assemblage.

Table 6, which gives an over-all summary of the vein mineralogy, is based in part on the more detailed analysis of Table 7. The "percentage frequency" given for the principal metallic minerals is merely the percentage of 156 polished sections found to contain the specified mineral. To express the average quantities of each mineral (where present), "modal" percentages for each were averaged over those sections in which it was found, and the

TABLE 6. MINERALS OF THE TELLURIDE VEINS

	Mineral	Composition	Abundance % Frequency	Amount (Vol. %)	Hypo-gene	Super-gene
NATIVE ELEMENTS	*Gold	Au	27.6	7.7	X	X
SULFIDES	*Tellurium	Te	37.2	44.4	X	x?
	Chalcopyrite	CuFeS$_2$	48.0	3.4	X	
PRINCIPAL METALLIC MINERALS	*Galena	PbS	28.2	3.6	X	x
	*Marcasite	FeS$_2$	40.5	13.8	X	
	Pyrite	FeS$_2$	95.7	26.2	X	
	*Sphalerite	(Zn,Fe)S	56.4	5.7	X	
TELLURIDES	*Altaite	PbTe	38.4	13.5	X	
	*Calaverite	AuTe$_2$	26.3	15.8	X	
	*Coloradoite	HgTe	41.7	11.3	X	
	*Hessite	Ag$_2$Te	35.3	12.8	X	
	*Krennerite	(Au,Ag)Te$_2$	7.7	19.7	X	
	*Melonite	NiTe$_2$	40.4	4.1	X	
	*Petzite	AuAg$_3$Te$_2$	44.8	12.9	X	
	*Sylvanite	AuAgTe$_4$	59.6	20.4	X	
SULFOSALT	*Tetrahedrite	(Cu,Fe)$_{12}$Sb$_4$S$_{13}$	28.2	2.5	X	
OXIDES	*Goethite	FeO·OH	Common, trace to large amts.		x	X
	*Tellurite	TeO$_2$	Common, trace to moderate amts.			X
HALIDE	Fluorite	CaF$_2$	Locally common, moderate amts.		X	
CARBONATES	*Ankerite	Ca(Fe,Mg)(CO$_3$)$_2$	Very common, small to large amts.		X	X
PRINCIPAL GANGUE MINERALS	*Calcite	CaCO$_3$	Locally common, moderate amts.		X	
	*Dolomite	CaMg(CO$_3$)$_2$	Common, small to large amts.		X	
SULFATES	*Barite	BaSO$_4$	Locally common, trace to mod. amts.		x?	
	*Gypsum	CaSO$_4$	Common, small to moderate amts.		X	X
	*Jarosite	KFe$_3$(SO$_4$)$_2$(OH)$_6$	Common, trace to moderate amts.		X	X
SILICATES	Quartz	SiO$_2$	Ubiquitous, chief vein mineral		X	
	*Roscoelite	K$_2$(V,Al)$_4$(Si$_6$Al$_2$)O$_{20}$(OH)$_4$	Common, moderate amounts		X	
NATIVE ELEMENTS	Silver	Ag	Reported, trace to moderate amts.			X
	Mercury	Hg	Rare, trace amounts			X
	Amalgam	(Au,Hg)	Reported, rare, trace amts.			X
SULFIDES	*Argentite	Ag$_2$S	Rare, trace to minor amts.			X
	*Arsenopyrite	FeAsS	Rare, moderate amts.		X	
	*Bismuthinite	Bi$_2$S$_3$	Rare, minor amounts		X	
	Bornite	Cu$_5$FeS$_4$	Rare, trace amounts			X
	Bravoite	(Fe,Ni)S$_2$	Rare, trace amounts		X	

OTHER MINOR AND RARE VEIN MINERALS

Mineral	Formula	Occurrence		
*Chalcocite	Cu_2S	Common, very minor amts.		X
Cinnabar	HgS	Reported, rare, small amts.	X?	X
Covellite	CuS	Rare, trace amounts	X	
*Molybdenite	MoS_2	Rare, trace amounts	X?	
*Stromeyerite	$AgCuS$	Rare, trace amounts	X	
TELLURIDES				
*Empressite	$AgTe$	Very rare, minor amounts	X	
*Nagyagite	$Au(Pb,Sb,Fe)_8(S,Te)_{11}$	Rare, trace amounts		X
Rickardite	Cu_3Te_2	Rare, trace to minor amounts	X	
*Stuetzite	Ag_5Te_3	Rare, trace to minor amounts	X	
*Tetradymite	Bi_2Te_2S	Uncommon, minor amounts		
Weissite (?)	Cu_2Te	Very rare, trace amounts		X
SULFOSALTS				
*Aikinite	$PbCuBiS_3$	Very rare, minute traces	X	
Kobellite	$Pb_2(Bi,Sb)_2S_5$	Reported, rare traces ?	X?	
*Jamesonite	$Pb_4FeSb_6S_{14}$	Rare, trace amounts	X	
*Pyrargyrite	Ag_3SbS_3	Rare, trace to minor amounts	x	
OXIDES				
*Hematite	Fe_2O_3	Uncommon minor amounts (specularite rare)		X
Ilsemannite	$Mo_9O_8 \cdot nH_2O(?)$	Reported, common locally, minor amounts ???		X
*Paratellurite	TeO_2	Rare, small amounts		X
*Pyrolusite	MnO_2	Very rare, minor amounts		X
HALIDES				
*Calomel	$HgCl$	Very rare, trace amounts		X
Cerargyrite	$AgCl$	Reported, rare small amounts		X
"Embolite"	$Ag(Cl,Br)$	Reported, rare small amounts		X
Iodyrite	AgI	Reported, rare small amounts		X
CARBONATES				
*Rhodochrosite	$MnCO_2$	Rare, locally small amounts	X	
*Siderite	$FeCO_3$	Uncommon, small amounts	X	
TELLURATE Montanite	$(BiO)_3(TeO)_4 \cdot 2H_2O$	Reported, local trace amounts		X
PHOSPHATE Woodhouseite	$CaAl_3(PO_4)(SO_4)(OH)_6$	Rare, small amounts	X	
TUNGSTATES *Ferberite	$FeWO_4$	Locally common, trace to large amts.	X	
Scheelite	$CaWO_4$	Reported, very locally present in moderate (?) amounts	X	
SILICATES Adularia	$KAlSi_3O_8$	Common, trace to minor amounts	X	
Chalcedony	SiO_2	Rare, small amounts	X?	
Opal	$SiO_2 \cdot nH_2O$	Rare, moderate amounts	X?	X?

WALL ROCK ALTERATION PRODUCTS — Hydromica, kaolinite, montmorillonite, orthoclase, quartz, sericite, siderite

MINERALS OF THE LEAD-SILVER STAGE ASSOCIATED WITH TELLURIDE ORES IN PLACES — Chalcopyrite, galena, gold, miargyrite, polybasite, proustite, pyrargyrite, sphalerite, stephanite, stromeyerite, tetrahedrite

*Confirmed by X-ray analysis.

TABLE 7. ESTIMATED MINERAL ABUNDANCES IN INDIVIDUAL TELLURIDE-BEARING POLISHED SECTIONS
(IN VOLUMETRIC PERCENTAGES OF METALLICS)

MINE	DISTRICT	SAMPLE NO.	Altaite	Calaverite	Chalcopyrite	Coloradoite	Galena	Gold	Hessite	Krennerite	Marcasite	Melonite	Petzite	Pyrite	Sphalerite	Sylvanite	Tellurium	Tetrahedrite	Others
Alpine Horn	G	18	10		tr	4	1			10			10	tr	3	61		tr	
"	G	26	8	20								2		30			40		
"	G	153	1		4											20	76		
"	G	158	30			4	tr					5		1	tr	25	69		
American	M	3				2	36		tr		2			56	50	4			
Black Rose	G	53			10	38		tr			36	5		1	2				
"	J	25			tr	2	3		2		2		2	5		3		2	
"	J	32	tr		45		tr		25				44	33	5	2		5	Nagyagite = 1, Stuetzite = tr
"	J	162	3		2	5						10		20		8	40	2	
Bondholder	J	95		60	1		tr	1			tr		5	48	2				Hematite = 4, Weissite? = tr
Buena ("Wano")	J	13			tr		tr	tr	2					15		40			
"	J	20			1	tr			3		7		20	24					
"	J	97	tr		2							tr	5	94	tr	40			
"	J	102			tr						6			59	tr	2			
"	J	106a		45	3						10	tr		65	1	3			
"	J	106	45			10	tr	1			2			24		30			
"	J	321b			tr	37	3	tr	60		49			5	3	20			
Bumble Bee	G	436			tr				90		tr	3		1	2	2			Nagyagite = 2, Rickardite = tr
Cash	G	32	7		tr	8	2					3		19	1	30			
"	G	150		13								1		5		40	15	tr	
"	G	151			tr						tr			6		7	69	tr	Stuetzite = 3
Cold Spring-Red Cloud	G	86			3		5	1	17			3	17	60					Aikinite = 2, Tetradymite = 2
"	G	88			1		5	2	20				29	40					Tetradymite = 6, Aikinite = 3
"	G	89							20			14	10	45		5			Aikinite = 3
"	G	111			tr			4	40				25	24	tr				Aikinite = 3, Tetradymite = 3
Colorado	G	42			4			10	12					70	3	2		1	Tetradymite = 2, Pyrargyrite = 3
"	G	43			1			2	4					73	5			5	
"	G	44						tr	5		25			14	43			2	Argentite = 2, Chalcocite = 2

"	G	46	—	—	1	—	—	1	4	—	—	—	—	89	2	—	—	3	
"	G	57	—	—	2	—	—	20	20	—	—	—	—	50	4	—	—	2	Pyrargyrite = 2
"	G	58	—	—	2	—	—	5	15	—	—	—	—	70	4	—	—	2	Pyrargyrite = 2
"	G	59	—	tr	—	tr	—	1	2	—	—	—	tr	89	4	—	—	3	
"	G	60	—	—	2	—	—	3	3	—	10	—	3	67	7	—	—	3	Pyrargyrite = 2
"	G	61	—	—	tr	—	1	tr	—	tr	46	—	—	49	2	—	—	1	Pyrargyrite = tr
Croesus	G	120b	1	—	—	2	—	—	—	—	—	5	—	10	—	62	20	—	
Eclipse	M	87	—	—	—	—	—	—	—	—	15	30	—	24	tr	—	30	—	Hematite = tr
"	M	9361	—	—	—	—	—	—	tr	—	5	7	—	10	tr	2	75	—	
"	M	9362	—	—	—	—	—	—	tr	—	20	6	—	19	2	1	51	—	
Emancipation	G	123	—	3	—	—	—	30	—	—	50	—	2	10	5	—	—	—	
"	G	141	—	12	8	—	4	1	—	—	35	—	4	27	8	—	—	1	
"	G	152	—	4	—	9	—	—	—	—	2	4	—	28	—	—	33	—	Tetradymite = 20
"	G	155	—	10	—	20	—	—	—	—	3	5	—	27	—	—	35	—	
"	G	156	—	13	—	20	—	—	—	—	1	6	—	5	—	—	55	—	
Empress	G	300	—	—	—	30	—	—	—	—	—	15	—	30	—	10	—	5	{Empressite = 4 {Stuetzite = 6
Fourth of July	J	10834	70	—	3	—	tr	—	—	—	tr	—	—	2	4	20	—	tr	
Franklin	G	131c	—	—	—	5	—	—	—	—	—	3	—	8	—	55	29	—	
"	G	131d	—	—	—	—	—	—	—	—	—	—	tr	5	—	94	—	—	
Freiberg	G	110	—	—	—	—	—	4	49	—	—	—	35	4	—	tr	—	tr	Tetradymite = 7
Gladiator	J	11	—	—	15	—	—	—	1	—	—	—	—	74	5	—	—	5	
"	J	103	2	—	7	—	tr	—	31	—	—	5	30	14	3	2	tr	5	
"	J	132	—	—	2	—	tr	—	38	—	—	1	54	tr	2	1	—	1	
Golden Age	J	98	tr	—	3	—	—	—	—	52	—	—	6	3	34	1	—	—	
" "	J	10734	2	—	3	—	5	—	—	33	—	—	2	10	40	5	—	—	
Golden Harp	G	140	3	—	—	33	9	—	—	—	25	5	—	24	tr	—	—	—	
Grandview	G	120b	tr	—	—	1	—	—	—	40	—	5	—	3	—	11	40	—	
"	G	134	tr	—	—	2	—	—	—	—	—	—	—	6	—	1	—	—	Ferberite = 90
"	G	135b	1	—	—	6	—	—	—	—	—	4	—	3	—	36	50	—	
Gray Copper	J	116a	—	—	tr	—	7	7	5	—	tr	—	—	35	40	—	—	3	Pyrargyrite = 3
Grouse	J	104	—	—	tr	—	—	—	4	—	—	—	8	79	—	8	—	—	
"	J	142	2	1	—	—	—	3	—	—	—	—	2	90	—	1	—	—	Weissite? = 1
Herald	G	118	—	1	—	4	—	2	—	—	20	—	1	60	2	—	—	—	Ferberite = 10
"	G	119	—	1	—	7	—	3	—	—	tr	—	3	69	2	—	—	—	Ferberite = 14
Horsefal	G	137	12	20	1	—	tr	—	3	—	—	2	15	38	3	3	—	tr	{Weissite? = tr {Rickardite = 1
Ingram	G	16	—	—	tr	—	2	—	8	—	—	tr	3	32	2	50	—	2	Pyrargyrite = tr
"	G	37	—	—	—	—	—	—	3	—	—	tr	6	19	—	71	—	tr	
"	G	39	—	—	—	2	—	—	—	—	—	3	5	30	—	60	—	—	
"	G	40	—	—	tr	—	—	—	5	—	—	tr	5	9	—	80	—	tr	

TABLE 7 (CONTINUED). ESTIMATED MINERAL ABUNDANCES IN INDIVIDUAL TELLURIDE-BEARING POLISHED SECTIONS (IN VOLUMETRIC PERCENTAGES OF METALLICS)

MINE	DISTRICT	SAMPLE NO.	Altaite	Calaverite	Chalcopyrite	Coloradoite	Galena	Gold	Hessite	Krennerite	Marcasite	Melonite	Petzite	Pyrite	Sphalerite	Sylvanite	Tellurium	Tetrahedrite	Others
Ingram	G	107	2	1	1	—	—	—	10	—	—	3	5	40	1	40	—	—	
"	G	1136	4	10	—	1	—	—	—	—	—	1	2	83	—	10	—	—	
John Jay	J	1	tr	10	—	tr	—	—	—	—	5	2	—	24	1	—	53	—	
"	J	8	—	10	tr	20	tr	—	—	—	—	5	—	34	tr	1	30	—	
"	J	15a	—	30	tr	8	—	—	—	—	—	—	—	9	—	1	50	—	
"	J	15	1	35	tr	5	—	—	—	—	3	3	—	9	2	5	44	—	
"	J	49	3	—	—	7	—	—	—	—	2	—	—	20	2	10	62	—	
"	J	50	30	—	—	4	—	—	—	—	40	—	—	8	—	14	73	—	
"	J	56	—	13	—	3	tr	—	—	—	—	1	—	25	tr	1	27	—	
"	J	66	1	10	—	5	—	—	—	—	—	2	—	9	—	—	30	—	
"	J	69	8	10	—	13	—	—	—	—	—	6	—	55	1	2	20	—	
"	J	148	—	—	—	15	—	—	—	20	—	1	—	29	—	6	35	—	
"	J	7532	25	15	—	4	—	—	—	15	9	tr	—	20	1	20	61	—	
"	J	75c32	tr	—	—	—	—	—	—	—	—	—	—	6	—	—	50	—	
"	J	75321	68	—	—	2	—	—	—	—	2	1	—	15	tr	—	37	—	
"	J	75322	—	—	—	—	—	—	—	—	4	tr	—	39	tr	—	45	—	
"	J	75h32	—	—	—	15	—	—	—	—	—	—	—	3	tr	10	—	—	
"	J	157	—	—	—	—	—	tr	2	—	—	1	4	29	tr	3	63	2	
Kekionga	M	2TL69	2	30	—	tr	—	—	—	—	2	tr	—	5	—	tr	—	1	Ferberite = 86
Keystone	M	27	—	25	5	—	—	—	2	—	4	2	36	19	5	—	50	tr	Molybdenite = 3
King Wilhelm	J	108	—	—	15	—	5	10	2	—	—	—	8	15	5	10	—	—	Nagyagite = 3
"	J	3d33	2	—	10	—	tr	—	tr	—	—	—	32	50	5	1	—	—	Nagyagite = 4
"	J	3c331	—	—	—	—	tr	10	3	—	—	—	—	39	—	—	—	—	
"	J	3c332	—	—	tr	—	—	—	—	—	—	—	20	2	34	—	—	4	{ Mineral B = 6 / Mineral B = 25
Lady Franklin	M	94	1	5	tr	21	—	—	—	—	15	5	—	10	1	3	40	—	
"	M	109	tr	1	tr	20	—	—	—	—	—	—	—	28	—	8	42	—	
"	M	10-36	1	—	—	15	—	—	—	—	4	3	—	19	1	3	58	—	
Last Chance	J	38	tr	—	—	tr	—	7	—	—	—	—	—	15	—	5	74	—	Rickardite = 5
"	J	5-32	tr	—	—	83	—	—	5	—	—	—	—	—	—	—	—	—	{ Mineral B = 5 / Nagyagite = 2
Little Johnny	G	126	2	10	—	—	—	1	—	—	10	—	5	60	5	—	—	—	

Locality			Nagyagite = 5 / Nagyagite = 2, Pyrargyrite = 2 / Bravoite = 2 / Molybdenite = 2 / Mercury = 5, Weissite? = tr / Rickardite = tr / Rickardite = 4, Weissite? = 2
Logan	G	30	
"	G	35	
Melvina	G	28	
"	G	145	
Monitor Tunnel	J	154	
Nancy	G	3–36	Nagyagite = 5
"	G	9	{ Nagyagite = 2, Pyrargyrite = 2
"	G	19	
	G	41	
"	G	54	
New Era	G	159	
"	G	163	
New Rival	G	113	Bravoite = 2
"	G	114	
Osceola-Interocean	J	163	
Plough Boy	G	13–36	Molybdenite = 2
Poorman	G	130	
"	G	105	
"	G	17	{ Mercury = 5, Weissite? = tr
"	G	31	
"	G	47	
"	G	51	
"	G	64	
"	G	96	
"	G	122	Rickardite = tr
"	G	139	
"	G	124	
"	G	736	{ Rickardite = 4, Weissite? = 2
Potato Patch	G	7	
Richmond	G	24	
"	G	34	
Shirley	E	22	
"	E	1436	
Slide	G	10	
"	G	128	
"	G	143	
Smuggler	J	6	
"	J	29	
"	J	48	

TABLE 7 (CONTINUED). ESTIMATED MINERAL ABUNDANCES IN INDIVIDUAL TELLURIDE-BEARING POLISHED SECTIONS
(IN VOLUMETRIC PERCENTAGES OF METALLICS)

MINE	DISTRICT	SAMPLE NO.	Altaite	Calaverite	Chalcopyrite	Coloradoite	Galena	Gold	Hessite	Krennerite	Marcasite	Melonite	Petzite	Pyrite	Sphalerite	Sylvanite	Tellurium	Tetrahedrite	Others
Smuggler	J	62	39		2		tr	tr	5		5		42	tr	1	10		tr	
"	J	65	50	3	1	15		10			1		15	2	2				
"	J	67	57	tr	tr		3	15					15	5	5			tr	
"	J	68	59		1		1	10					19	1	3				
"	J	120	40			4	2			2?			15		15	26			
"	J	1361	64				tr	12	5				15	7	5	3			
"	J	1362	4		2								30		5	50			
"	J	332	51			2						1	30		2	10		1	
"	J	JT5	3			3								5		50	36		Stuetzite = 4
"	J	JT6	5											3		38	45		Stuetzite = 3
Sparkling Jewell	G	139-a												97	2	1	2		
Sparkling Jewell	G	139-b		tr										97		tr	2		
Stanley	J	100						6						34					
"	J	236						1		20			tr	10					
Sterling	G	127										2		29					Bismuthinite = 35 / Tetradymite = 24
White Crow	G	4			1	4						15		5		2	48		Bismuthinite = 89
"	G	5			2	3						15		20		2	73		
"	G	21		1							70			14			60		
Winona	G	136			1			tr	56			15	3	24			15		

G = Gold Hill   J = Jamestown   M = Magnolia   E = Eldora resulting figures are given in Table 6 under "Amount (Vol. percentage)." The 16 minerals for which such data are given actually comprise 95 percent (by volume) of all metallic assemblages in the samples analyzed.

In the following pages, each mineral is described approximately in the order of J. D. Dana (E. S. Dana, 1892). Emphasis is given to the most abundant vein minerals and to those of greater mineralogical interest or genetic significance. A thorough and accurate review of literature on mineral occurrences in Boulder County has been prepared by Eckel (1961) who gives the history of discovery and subsequent record for many of the important vein minerals dealt with in the present paper.

NATIVE ELEMENTS

Gold

Native gold, the chief ore mineral in the oxide zone, is also a very important source of values in the hypogene telluride ores. In the primary ores, it occurs in amounts subordinate to several of the gold-bearing tellurides, but in those samples studied it accounts for approximately 42 weight percent of the total gold. About 28 percent of the polished sections contain some hypogene gold in amounts averaging 7.7 percent of the metallic assemblages.

The early mining records carry no estimates of the relative importance of hypogene and supergene gold, but from available assay data, known mining depths, and assumed average depths of oxidation, it is estimated that supergene gold accounted for roughly 5 percent of total production in the Jamestown and Gold Hill districts. In the Magnolia district, where average mining depths were more shallow, oxide gold probably accounted for about 15 percent of the total production.

Supergene gold is normally found within 100 feet of the surface, but may occur at depths up to 200 feet where oxidation is unusually deep. It is abundant only in leached cappings within 5 to 60 feet of the ground surface.

Gold derived from the tellurides is typically very fine grained and is embedded either in iron or tellurium oxides. Most of the gold mined from the vein outcrops was finely dispersed through spongy limonite forming what the miners called "rusty gold." In polished sections, disseminations of fine gold in limonite or tellurite have commonly replaced the gold-bearing tellurides (Pls. 2a, 6c and 9c).

Supergene intergrowths of gold and rickardite were formed as early alteration products of the gold tellurides and are similar to those described at Kalgoorlie by Stillwell (1931). The gold forms a network of veinlets and very fine filaments in the copper telluride, and is in places so abundant that it actually predominates in the intergrowth. In ores of the Potato Patch, masses of intergrown gold and rickardite are enclosed by tellurite and mark the sites of original grains of calaverite in native tellurium (Pl. 12e). Similarly, in one sample from the Last Chance mine, inclusions of sylvanite in coloradoite have altered to rickardite which is cut by myriad veinlets of free gold (Pl. 9d). Some gold has migrated into the coloradoite where it forms distinct replacement veinlets.

Wire gold is fairly common in oxidized ores where it is intergrown with limonite and other oxidation products. The wires are typically 0.3 mm in diameter, several millimeters long, and curved or tightly coiled. Abundant wire gold of the Cold Spring-Red Cloud ores appears in oxide crusts on partially decomposed hessite and petzite (Pl. 7d). In this occurrence, some of the wires are blackened by films of manganese oxides that can be flaked off, revealing the bright gold beneath. Although wire gold is most common near the surface, it occurs in oxidized ores as deep as 150 feet in the Emancipation mine.

Where limonite or tellurite is lacking, and in samples from shallow mine workings, it is difficult to distinguish supergene from late hypogene gold. This is the case at the Colorado and nearby Cash mines of Gold Hill where the evidence seems to indicate that some gold may have been deposited from descending surface waters. In samples taken in 1959 along the tunnel level of the Colorado vein 42 feet below the surface, gold occurs as thin films along small irregular fractures that crosscut the banded, sulfide-rich telluride ore (Pls. 4f and 13c). Along these fractures the gold coats and follows the irregular traces of the sulfides, particularly pyrite, in the vein. The hypogene sulfide layers evidently served as precipitants of gold migrating along the small fractures close to the surface. Normal hypogene gold occurs *within* the vein structure where it replaces sphalerite, tetrahedrite, pyrargyrite, and hessite. This gold does not replace pyrite and yet it is pyrite which most conspicuously localizes gold along the small fractures. This difference, coupled with the fact that hessite closely associated with gold in the vein shows twinning normally attributed to inversion from its intermediate temperature form, suggests that both hypogene and supergene gold are present in this unusual ore. Almost identical specimens come from stope fill of the Slide mine but these specimens were not found in place.

Hypogene gold is also widespread in occurrence and, in contrast to supergene varieties, may be found at any depth extending to 1100 feet in larger mines of the Gold Hill district. Native gold and native tellurium are markedly antipathetic in occurrence; they may occur in the same veins but not in proximity, and hypogene gold and tellurium have never been found together in hand specimen or polished section. In some mines, notably the John Jay, free tellurium pervades the veins to the complete exclusion of hypogene free gold.

Sylvanite and gold also seem incompatible as they have never been seen in mutual contact and are rarely present together in polished section. In exceptional specimens from the Smuggler, these two minerals occupy the same vugs and come within a millimeter of touching, but even here are separated by areas of petzite.

Among the tellurides, gold is most commonly associated with hessite and petzite, less commonly with calaverite and coloradoite, and only rarely with sparse nagyagite and tetradymite. The strong preferential association of gold and coloradoite characteristic of the Kalgoorlie deposits (Stillwell, 1931; Markham, 1960) is not developed in the Boulder ores despite the fact that both minerals are abundant. Of interest is the fact that altaite and gold, a pair rarely found in direct contact, comprise a predominant association in ore of the Smuggler mine at the northern extremity of the area of telluride mineralization. The gold veins both altaite and petzite, and is the latest ore mineral (Pl. 2b). Native gold has not been found in direct contact with stuetzite, empressite, krennerite, or melonite.

Gold was clearly the latest metallic mineral to form in the complex depositional sequence. In most ore it formed by replacement of earlier ore minerals rather than by filling of available open spaces. It is found as irregular veinlets in all of the tellurides with which it is commonly associated (Pl. 2c).

In comparison with ores containing free tellurium, those bearing native gold are noticeably richer in sulfides and sulfosalts. In many specimens, marcasite, sphalerite, galena, chalcopyrite, and tetrahedrite are veined by gold and its common associates, hessite and petzite. In some ores, gold is interstitial to fine-grained pyrite and has selectively replaced original sphalerite and tetrahedrite (Pl. 2d).

Roscoelite, a fairly common gangue mineral in the telluride veins, is preferentially associated with native gold and the gold-bearing tellurides to the extent that green, roscoelite-bearing quartz was sought by the miners as a guide to ore. The roscoelite forms discrete layers or intergrowths with quartz and is veined and replaced by the younger tellurides and free gold (Pl. 13b).

In a number of high-grade specimens from the Croesus, Emancipation, Gladys, Golden Harp, New Era, and Slide mines, native gold is the major metallic mineral; the gold formed in open spaces and, unlike most of its occurrences, is here accompanied by very minor amounts of the tellurides or none at all. The gold appears either as small irregular grains scattered uniformly through porous vein quartz (Pl. 2e) or as gold-ankerite vein fillings. In the gold-carbonate intergrowths the gold forms leaves and plates measuring from 0.2 by 0.7 mm to 0.2 by 3.0 mm in section. These appear to have grown on older layers of quartz that border the veins and were subsequently enveloped by younger ankerite. This age relationship is best revealed in some specimens where the leaves of gold extend through ankerite and into open spaces left unfilled by the carbonate. The quartz along the borders of these veinlets is fine grained (1 mm), subhedral and includes roscoelite, pyrite, marcasite, sphalerite, and galena. The width of the veinlets ranges from 3 to 12 mm. Minor petzite and calaverite occur in this type of gold ore (e.g., Emancipation mine) and form small discrete patches within the ankerite or at the edges of the gold leaves. Leaf gold of this type has been found on the 1100-foot level of the Slide mine and thus is thought to be of hypogene origin.

There are no available chemical analyses of gold from the telluride veins but its rich yellow color suggests that it is all at least 900 fine. There is no discernible variation in the color of gold in its different hypogene associations and no conspicuous color difference between the hypogene and supergene varieties.

Tellurium

As pointed out by Eckel (1961, pp. 320-321), native tellurium was identified in ores of the Red Cloud mine, Gold Hill district, soon after the initial discovery of gold telluride ore on that property in 1872 (Endlich, 1874; Silliman, 1874a, b). Shortly thereafter, abundant tellurium was found in the Keystone, Mountain Lion, Dun Raven, and Smuggler mines and nearly pure specimens of tellurium weighing up to 25 pounds were extracted from the John Jay mine in the Jamestown district (Genth, 1877; Jennings, 1877). Subsequent studies have verified the common occurrence of the mineral in the telluride veins (Wilkerson, 1939; Headden, 1903; Ives, 1935; Lovering and Goddard, 1950; Wahlstrom, 1950).

The present analysis confirms the abundant and widespread distribution of native tellurium in the telluride veins and an additional 20 mines can now be added to the occurrences previously cited in the literature (see Table 7). From the polished section data it is estimated that approximately one fourth of all the tellurium in the veins is uncombined. The mineral was recorded in 58 of the 195 polished sections examined. In view of the abundance of uncombined gold in these same deposits, it is surprising that free tellurium occurs in such quantity. Of course, as previously noted, hypogene gold and tellurium are rarely abundant in the same veins and are never together in hand specimen or polished section.

The abundance of native tellurium explains in large part the disappointing assay returns frequently received by the miners for ores which appeared to be high grade but which probably contained substantial native tellurium and little of the gold-bearing tellurides.

The numerous tellurium occurrences can be grouped as (1) euhedral to subhedral crystals lodged in open, quartz-lined seams or vugs, or embedded in fine-grained vein quartz, (2) anhedral, polycrystalline aggregates complexly intergrown with tellurides and enclosed by vein quartz, and (3) a special variety of very fine-grained tellurium which makes up a porous, spongy aggregate found replacing older tellurides in some places. The porous variety, here called "sponge tellurium," has not been previously described.

Crystal forms are best displayed by tellurium that grew in open seams but did not completely fill them, and it is such spectacular occurrences that are best described in the early literature. The forms reported include (1) small hexagonal cleavage plates, (2) hexagonal prisms with perfect prismatic and poor basal cleavage, and (3) very thin, wafer-like prisms intergrown with the tellurides. Individual crystals are typically 1 to 2 mm in longest dimension, but prisms up to 2 cm long occur in John Jay samples and Endlich (1874) described crystals somewhat larger than this from the Cold Spring-Red Cloud mine. Prisms are commonly striated or grooved parallel to the c-axis and many are curved or bent. Genth (1877) described small, distorted crystals in the John Jay ores that have a deeply grooved and cavernous appearance and the writers have observed the same sort of crystals from that property, some having a fused and skeletal appearance. In ore from the Buena mine, single tellurium crystals measuring 0.5 by 2.0 mm are associated with small 1.0 by 4.0 mm elongate crystals of sylvanite. Both minerals occupy cavities in an older intergrowth of quartz, roscoelite, and fluorite, and are in places coated by thin crusts of microcrystalline quartz.

Some tellurium crystals that formed in open vugs were evidently attacked by later ore solutions with the development of outer films and coatings of younger tellurides. This relationship is rarely seen in polished section due to plucking of the crystals, but it is clearly demonstrated in ores from the Emancipation mine (Pl. 2f). Here, the tellurium is marginally replaced by coloradoite, sylvanite, and tetradymite and the outlines of the original tellurium crystals are still evident. Such occurrences are significant proof that at least some tellurium is older than the associated tellurides.

Where tellurium grew in competition for space with contemporaneous vein quartz, its prismatic forms are still evident but poorly developed (Pls. 2g and 13a). Elongate, subhedral crystals of tellurium embedded in quartz are very commonly bordered by rims of altaite, coloradoite, sylvanite or calaverite, and melonite (e.g., Pls. 5a and 5h), or more rarely stuetzite (Pl. 9g). This texture is well illustrated by Wahlstrom (1950) who interprets the rims of the type shown in Plates 7h and 8b as early incrustations on the walls of vugs subsequently filled by late native tellurium. Some such intergrowths may form in this way, but the explanation cannot apply to occurrences of the Emancipation type (Pl. 2f) where there are no supporting vug walls, nor to textures in which both the rimming tellurides and the surrounding vein quartz conform to the elongate morphology of the tellurium crystals (Pl. 7g). Melonite is particularly common as a rim mineral around tellurium, and can be seen in polished section as bands that follow the outlines of tellurium crystals (Pls. 7g and 7h) or as fine-grained aggregates pseudomorphous after bladed tellurium (Pl. 8a).

A large proportion of native tellurium in the veins occurs as anhedral polycrystalline aggregates intergrown with the tellurides and filling vugs in vein quartz. Textures in this type of ore are more complex and age relationships correspondingly obscure. The associated tellurides occur in places as inclusions within the tellurium or are concentrated along the margins of tellurium aggregates forming the rims described above. In many specimens, tellurium is mutually intergrown with coloradoite, sylvanite, or calaverite, and the minerals appear contemporaneous (Pl. 5d). In other places, however, late anhedral tellurium appears to have filled in around euhedral or subhedral crystals of calaverite or sylvanite that project from quartz walls into vugs (Pls. 5c and 10e). In some polished sections, native tellurium appears to have veined and partially replaced older vug fillings of coarse calaverite, krennerite, or sylvanite (see Pl. 2h).

A comparatively rare but significant type of tellurium is the variety here called "sponge tellurium" which is represented in ores of the Bumble Bee, Emancipation, and Lady Franklin mines. It forms late, porous, fine-grained aggregates that occur as marginal replacements of sylvanite, tetradymite, coloradoite, and melonite (Pls. 3a, 3b and 11c). The material gives an excellent X-ray powder pattern for tellurium. The porosity of this tellurium bears no obvious relationship to the tellurides that have been replaced. Sponge tellurium and normal, coarsely crystalline tellurium occur locally in close proximity but do not grade into one another. In ores of the Emancipation mine, tetradymite replaces early tellurium and is in turn selectively replaced by sponge tellurium (Pl. 3b). This variety of tellurium may originate either through late hypogene leaching of the tellurides or possibly as a supergene product. It has not been found in association with limonite, tellurite, or other known supergene minerals and, unfortunately, its vertical distribution in the veins is unknown.

Aside from sponge tellurium which may be supergene, most of the tellurium observed is probably hypogene. It is intergrown with other primary ore minerals over the full vertical mining range and in many specimens appears to be older than many of the tellurides with which it is associated. Most of the specimens containing tellurium are made up of fresh metallic assemblages that show no signs of supergene alteration. An exception may be represented by some samples from shallow workings of the John Jay mine in which crystals of tellurium, free of the usual association with tellurides, occupy quartz-lined seams in altered granite. This tellurium is associated only with tellurite that stains the seams a conspicuous yellow.

An unusual deformational texture is seen in tellurium of the White Crow mine (Pl. 3c). An original intergrowth of tellurium, melonite, coloradoite, and sylvanite in quartz was deformed with the result that angular fragments of quartz and the tellurides are strung out in tellurium which evidently responded to stress more plastically. The individual grains of tellurium show undulatory extinction. Minor late chalcopyrite veins the tellurides, but not the unfractured tellurium.

Tellurium is most commonly associated with altaite, calaverite, coloradoite, marcasite, melonite, pyrite, sphalerite, and sylvanite but is also found in contact with chalcopyrite, krennerite, stuetzite, ferberite, and tetradymite. Its marked antipathy to gold has been stressed and it has not been seen in contact with hessite, petzite, or galena. Samples that contain native tellurium contain appreciably less chalcopyrite, sphalerite, galena, and tetrahedrite than do ores bearing free gold.

Mercury and Amalgam

Sparse, noncommercial quantities of mercury and amalgam occur as oxidation products of coloradoite in near-surface parts of the telluride veins. Eckel (1961, pp. 42, 220) has reviewed previous reports of these minerals in Boulder County where scattered occurrences have been noted in the

Jamestown, Gold Hill, and Magnolia districts (Endlich, 1874, 1878; Smith, 1883; Lovering and Tweto, 1953). These rare minerals are difficult to find in the oxidized ores, but if a suspect sample is gently pounded, the mercury present is driven to open spaces where it coalesces into visible globules (Lovering and Tweto, 1953, pp. 35-36). Using this procedure, the writers have found mercury in dump samples from the Poorman mine where it is associated with tellurite, minute traces of calomel, and relict coloradoite. Microchemical tests revealed no silver in this mercury from the Poorman. Previous reports of amalgam have not been confirmed, but its occurrence in minor amounts seems reasonable and well established.

Silver

Native silver was reportedly found as an oxidation product in some of the telluride veins (Endlich, 1874, 1878), but the writers have comparatively few good specimens of the near-surface ores and have been unable to confirm this species. Wires and leaves of native silver were common in near-surface ores of the Cold Spring-Red Cloud, but were not present in the few rich oxidized samples we have from that property. What at first was thought to be wire silver in these specimens proved to be gold blackened by brittle surface films of manganese oxide.

SULFIDES

Chalcopyrite

Small, but widespread quantities of chalcopyrite were deposited during the telluride stage of mineralization, commonly in close association with sphalerite, tetrahedrite, and galena. It is a major constituent of some ores from the Black Rose, Golden Age (Sentinel vein), Gladiator, and Temborine mines, but it more typically occurs in amounts subordinate to the associated tellurides and native metals. About half of the telluride ores examined contained chalcopyrite in amounts averaging 3.4 percent of the total metallic assemblages.

Most chalcopyrite formed as a late sulfide but is typically older than the tellurides and native metals (Pl. 3f). It is commonly molded against early pyrite and occurs as a replacement product after marcasite, sphalerite, tetrahedrite (Pl. 3d), and rare molybdenite. Chalcopyrite and galena show mutual textures and reversed veining relationships in different specimens, but mostly chalcopyrite appears to be the younger mineral.

In places, deposition of chalcopyrite overlapped the early tellurides and small amounts of chalcopyrite locally vein sylvanite (e.g., Bondholder, White Crow mines, Pl. 3e), or are intergrown with petzite in veinlets transecting older calaverite, sylvanite, or krennerite. Typically, however, this sulfide predates the tellurides and is veined or replaced by them (Pl. 3f).

A minor but interesting occurrence of chalcopyrite is represented by ores of the Freiberg and Cold Spring-Red Cloud mines where it seems to have formed as a by-product in a complex replacement reaction. Under the microscope it appears as minute blebs intergrown with laths of tetradymite and both minerals form a reaction rim along the contacts of aikinite ($PbCuBiS_3$) with an intergrowth of hessite and petzite (Pl. 11d). The chalcopyrite and tetradymite probably formed as aikinite replaced an unstable silver telluride (the "x phase" discussed in later pages) which subsequently broke down upon cooling to form the present hessite-petzite intergrowths.

It is noteworthy that copper in the telluride veins was deposited chiefly as chalcopyrite and tetrahedrite early during the telluride mineralization and does not itself form a telluride. Careful search has failed to reveal possible hypogene rickardite or vulcanite. In the Boulder ores, copper tellurides formed only as rare supergene products.

Galena

An important ore mineral of the lead-silver veins, galena is only a minor accessory in the telluride deposits. Twenty-eight percent of the samples examined carry this mineral in amounts averaging 3.6 percent of the metallic assemblages.

The textures and age relationships of galena show little variation in the ores. It is molded against or corrodes older pyrite, and also veins associated sphalerite and tetrahedrite. In a majority of cases, it is veined by chalcopyrite, but in many places the two minerals are mutually intergrown or galena veins chalcopyrite. In Lady Franklin ore, galena appears as scalloped inclusions in chalcopyrite producing the classical "caries texture." Galena of the Poorman mine in places forms replacement rims around older sphalerite and replacement veinlets that cut both sphalerite and tetrahedrite (Pl. 11e).

Galena is veined by hessite, petzite, and gold (Pl. 2c) and appears as irregular or round relicts in many of the common tellurides. Corroded patches of galena are particularly common along the walls of vugs and seams filled by the younger tellurides. In some ores, such as those of the Colorado vein, rounded inclusions of galena and sulfosalts are so numerous in hessite that they lend a mottled appearance to this mineral in polished section (Pl. 3g).

It is noteworthy that galena has never been found in direct contact with marcasite or sylvanite although it does accompany these minerals in some polished sections. Even more conspicuous is the fact that galena and native tellurium very rarely appear in the same polished sections and are never in direct contact in the ores.

Lead is one of the few metals that appears both as a sulfide and as a telluride in the telluride veins. In each case, it appears to have formed late in the particular sequence involved—as galena late in the sulfide sequence and later as altaite, the youngest of the tellurides. Galena and altaite locally occur in contact, but there is no especially close association between the two in the Boulder deposits.

Marcasite

Marcasite, an important accessory mineral, was found in 41 percent of the samples examined in amounts averaging 14 percent of the metallic minerals. It is very commonly intergrown with pyrite, and the details of this important association are considered in later pages.

As an early hypogene mineral, marcasite occurs (1) as prismatic crystals ranging from hair-like fibers several microns long to larger crystals up to 1 mm long which appear singly or as matted aggregates embedded in horn quartz (Pl. 4a), (2) as fibrous cores within pyrite (Pl. 6h) or fibrous overgrowths on pyrite, (3) as colloform spherical aggregates comprised of concentric fibrous layers and displaying syneresis (Pl. 4b); such aggregates are typically mounted on early quartz crystals and surrounded by younger, fine-grained horn quartz, (4) as small bladed crystals surrounded by younger tellurides and native metals in vuggy openings of horn quartz seams (Pl. 3h).

Another type of hypogene marcasite is represented in samples from the Bondholder, Osceola-Interocean, and Poorman mines. In these, marcasite appears as strongly anisotropic, sharply angular patches enclosed in coarse massive pyrite. Each area of marcasite displays sharp and uniform extinction, but the separate areas are of different shapes and orientation. The intergrowth gives the impression of an original mass of marcasite, parts of which inverted to pyrite after the initial deposition.

Marcasite occurs in contact with most of the common metallic minerals of the telluride veins. It is most commonly intergrown with pyrite because both iron sulfides are very abundant. Among the common metallic minerals, marcasite has not been found in direct contact with galena, krennerite, or tetrahedrite.

Deposited early in the vein sequence, hypogene marcasite formed along with early pyrite and is typically veined and replaced by the younger sulfides, tellurides and native metals with which it is associated.

One occurrence of clearly supergene marcasite was noted in a sample from the dump of the Poorman mine, but this was not in direct connection with a telluride vein. Here, bright metallic films of marcasite coat fractures in pyritized pegmatite. Pseudomorphs of hematite after original pyrite lie beneath the films of marcasite within the pegmatite. Aside from this one occurrence of secondary material, all of the marcasite observed in the telluride ores appears to be hypogene. It occurs over the full vertical mining range and its abundance does not relate in any way to the present topographic surface.

Pyrite

This mineral is by far the most abundant metallic species in the telluride veins occurring in over 95 percent of the polished sections studied in amounts averaging 26 percent of the metallic minerals. These figures actually underestimate pyrite abundance because samples were chosen for polishing which were as rich as possible in minerals other than pyrite or marcasite. Pyrite is abundant both as a vein mineral and as disseminations in wall rock immediately adjacent to the telluride veins.

Within the veins, pyrite is most common as fine-grained disseminations that lend a gray or black color to the telluride-bearing vein quartz. The pyrite is either uniformly distributed throughout the quartz, or concentrated in bands or zones that reflect variation in the precipitation of iron sulfide during more continuous deposition of quartz. Such pyrite is anhedral to pyritohedral or cubic and varies in size from 1 mm down to particles that can scarcely be resolved at highest magnifications. The average grain size is approximately 0.05 mm.

Vein pyrite also occurs (1) as coarse (1 to 10 mm) anhedral to euhedral grains or aggregates isolated in vein quartz or attached to the walls of vugs or seams within vein quartz, (2) as layers coating early vein quartz or altered wall rocks along the borders of telluride veinlets, (3) as concentric growths with atoll structure solidly encased in horn quartz, (4) as crusts on fragments of wall rock or breccia fragments of early vein quartz included and cemented by later vein quartz, and (5) rarely as minute cubes .05 mm in size dusted on free surfaces of late vein carbonates. In addition, a variety of intergrowths occur with marcasite and these are described below.

Pyrite in altered wall rock adjacent to the telluride veins is normally fine grained (less than 0.5 mm), and is either anhedral or forms distinct cubes or pyritohedra. Also, minute seams of pyrite less than 1 mm wide cut the altered wall rock in places, and these are in turn crosscut by telluride veinlets. Pyrite also occurs in various stages of replacement of biotite in the wall rocks suggesting that at least a portion of pyrite in the walls formed through sulfidization of original mafic silicates.

Where pyrite is euhedral, both cubes and pyritohedra are common with the latter predominating. Some of the crystals show growth zoning which generally reveals that forms initially developed persisted throughout growth of individual crystals.

Pyrite occurs in contact with all metallic minerals known to be present in the veins. Most of it formed in the earliest stages of mineralization and is coated, veined, or replaced by the numerous ore minerals that followed in the vein sequence. Very rare vein specularite appears to be the only metallic mineral to precede pyrite. Although the bulk of iron sulfide formed in this early stage, very sparse quantities were again deposited in the closing stages of mineralization as evidenced by minute crystals seen coating late carbonates in ores of the Emancipation, New Era, and Poorman mines.

The association pyrite-marcasite is a particularly common one noted in 60 polished sections. These minerals appear separately or intergrown in either the wall rocks or vein matter or both. Their age relations differ from one specimen to the next but the record indicates that pyrite is most commonly the earlier mineral, is followed by intergrown marcasite-pyrite or marcasite alone, and then reappears once again to close the stage of iron sulfide deposition. Most individual specimens display only a part of this general sequence which becomes apparent only after tabulation of many observations. These minerals are irregularly intergrown in many samples, but in some they show an orderly arrangement of fibrous marcasite forming cores within or rims around pyrite grains (Pl. 6h). In some colloform marcasite growths, cubes of pyrite appear within the structure crosscutting the concentric layers (Fig. 9). Some pyrite aggregates have an elongate form suggestive of original blades of marcasite.

There is evidence that pyrite was somewhat more stable than marcasite in contact with solutions that deposited the tellurides and native metals. Where marcasite is in direct contact with these later minerals it shows some degree of replacement (Pl. 3h), whereas pyrite more commonly displays sharp crystal outlines against these same minerals (Pl. 7d). Euhedral forms of marcasite are best preserved where the mineral is encased in quartz (Pl. 4a) or otherwise shielded. In ores of the Nancy mine (Pls. 4c and 6h), thin (.006 to .03 mm) but persistent layers of pyrite separate marcasite from direct contact with the tellurides and appear to have armored the marcasite whose fibrous forms are exceptionally well preserved in this ore.

Sphalerite

Zinc has not been recovered in mining of the telluride ores, but sphalerite is present in accessory amounts in most of the veins, and locally, as in the

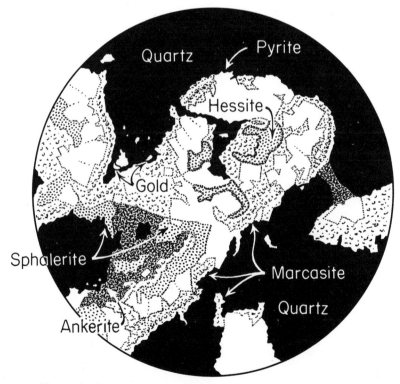

Figure 9. Tracing of photograph showing complex replacement of marcasite by younger vein minerals.

Cash and Colorado-Rex mines, it is a major constituent of the ores. About 56 percent of the polished specimens examined contain sphalerite in amounts averaging 6 percent of the metallic assemblages.

Most sphalerite formed early in the sulfide sequence but is typically younger than pyrite-marcasite and is veined and replaced by closely associated chalcopyrite, tetrahedrite, and galena (Pls. 3d and 11e). In an unusual sample from the Alpine Horn mine, sphalerite accompanies marcasite and pyrite in altered granite bordering telluride seams, but normally sphalerite occurs only within the vein. Marcasite veins sphalerite in samples from the Black Rose and Melvina, but these are reversals of typical marcasite-sphalerite age relations. The tellurides and native metals vein and replace sphalerite (Pl. 12b), and it is especially common to find irregular remnants of sphalerite in these later minerals or attached to the walls of vugs which they fill (Pl. 2b).

Some ores contain minor amounts of sphalerite that are atypically late; this sphalerite is light brown or yellow and presumably of lower iron content than the early type which is moderate brown and almost opaque in thin section. For example, telluride-bearing horn quartz of the American mine (Gold Hill district) is in places coated by an intergrowth of ankerite, pyrite, and light yellow-brown sphalerite. Also, in one sample from the Smuggler vein, curious scalloped layers of light brown sphalerite are draped over a rich quartz-telluride intergrowth.

Sphalerite of the Cash, Cold Spring-Red Cloud, Colorado, and Horsefal mines contain myriad blebs and filaments of chalcopyrite of the type commonly attributed to exsolution. Although chalcopyrite and sphalerite are intergrown in many other places, it is only in these ores that this particular intergrowth appears. It may therefore be significant that the mines involved are all in one restricted area close to the Hoosier reef near the town of Gold Hill.

Other Sulfides

The following sulfides occur in trace or minor quantities or are moderately abundant in very few places: argentite, arsenopyrite, bismuthinite, bornite, bravoite, chalcocite, cinnabar, covellite, molybdenite, and stromeyerite. Though quantitatively unimportant, these minerals lend variety and interest to the ores.

Arsenopyrite has not been found in direct contact with the tellurides but occurs in the Hawkeye vein of the Archer mine in horn quartz similar to that which carries tellurides elsewhere on the property. The arsenopyrite appears contemporaneous with most of the marcasite and pyrite with which it is intergrown but in some spots appears to vein marcasite.

Bismuthinite occurs in several unusual samples of telluride-bearing pyritic gold ore from the Stanley vein near Jamestown. This mineral forms slender blades several millimeters long intergrown with pyrite, tetradymite, and gold in coarse, glassy quartz. Originally thought to be sylvanite (Lovering and Goddard, 1950, p. 264), this bladed mineral gives a definite bismuthinite X-ray diffraction pattern.

Small amounts of bravoite occur in close association with melonite, marcasite, and pyrite in one sample from the Osceola-Interocean workings (Pl. 4d). The identification is based upon optical and etch properties as it has been impossible to extract sufficient quantities from polished section for X-ray analysis. The bravoite forms distinct crusts 10 to 20 microns thick on pyrite, marcasite, and quartz and in places separates these older minerals from vug fillings of melonite, hessite, and petzite. The bravoite displays pyritohedral and cubic forms against the tellurides but irregular contacts with quartz and the iron sulfides. Alongside pyrite, the bravoite is violet and its polishing hardness is between that of marcasite and melonite. Compositional zoning, so common in bravoite, is not displayed by this material. Melonite is in contact with the iron sulfides in many ores, but curiously enough this is the only known case where a nickel sulfide has developed along the $NiTe_2$-FeS_2 contacts.

Eckel (1961, p. 108) mentions a specimen in the collection of the Denver Museum of Natural History in which red specks of cinnabar coat the mercury telluride, coloradoite. The sample is from Boulder County but no specific locality is given. The writers have never found this mineral in the present suite and can only conclude that it is exceptionally rare in the telluride veins.

Unusual concentrations of molybdenite were found on lower levels of the Mountain Lion mine in the Magnolia district and several hundred pounds were extracted in the early mining days (Endlich, 1878). The precise relationship of these rich pockets to the telluride mineralization is, however, unknown (Lovering and Goddard, 1950, p. 233).

Lindgren (1907, pp. 454, 458) observed ore samples on mine dumps of the Eldora district which were stained blue by a mineral believed to be ilsemannite derived from molybdenite which he reported as abundant in the telluride deposits. The present authors have not seen many samples from the Eldora district, but in the main belt of telluride mineralization, molybdenite is a very minor constituent of the primary ores and no ilsemannite has been found in the oxidized ores. In the Poorman mine, sparse molybdenite is intergrown with pyrite in horn quartz that predates the tellurides (Pl. 4e). Traces of molybdenite also occur in some ore of the King Wilhelm mine; here, it appears to have formed after pyrite but before sphalerite. The molybdenite occurs as scattered flakes on the walls of vugs and is coated by sphalerite, embayed by chalcopyrite, and veined by petzite and native gold.

Stromeyerite is a common constituent of the lead-silver ores and as such is indirectly associated with the tellurides where both types of ore are present (e.g., Croesus and Yellow Pine mines). However, as a precipitate of the telluride stage of mineralization, this mineral is rare. Very minor amounts are intergrown with calcite in late, post-telluride vein fillings in samples from the Colorado vein (Pl. 4f).

Traces of argentite, bornite, covellite, and chalcocite have been recorded in various samples of partially oxidized ore from several mines. These occur as normal supergene alteration products of accessory sulfides that accompany the tellurides and are economically insignificant.

In many places, tellurides exposed in mine workings develop a black, greasy surface film which the miners call "tellurium grease," and which serves as a guide to the location of tellurides underground. The authors X-rayed scrapings of tellurium grease from ores of the John Jay and Gray Eagle mines fully expecting some silver telluride or tellurate but found that the material from both localities was actually argentite (acanthite), probably derived from decomposition of the silver-bearing tellurides.

TELLURIDES

Altaite
The telluride of lead occurs in more than a third of the polished sections examined in amounts averaging 13.5 percent of the metallic assemblages. It is therefore a widespread and abundant constituent of the ores though one of no economic significance.

Altaite is also the youngest of the tellurides, and occurs most commonly as replacement films and rims bordering the older tellurides and native tellurium. These relations are best shown by ores of the John Jay mine in which altaite forms bright, tin-white films on free surfaces of tellurium crystals and, in polished section, typically appears along the borders of tellurium-coloradoite-sylvanite vug fillings (Pls. 4g and 5a; Fig. 10). In many specimens, as in ores of the Eclipse mine, altaite is closely associated with melonite and both minerals occur in contact with quartz along the walls of cavities filled by intergrowths of tellurium with sylvanite or calaverite (Pls. 8b and 8g). Veinlets of altaite in the older sulfides are uncommon, and distinct veinlets of altaite in other tellurides are truly rare. Altaite in places veins petzite in some samples from the Slide mine (Pl. 5b).

Altaite is more abundant in assemblages containing native tellurium than in those bearing free gold, although it does occur in both types of ore. It has been observed in contact with all of the other common metallic minerals and is, to a moderate extent, preferentially associated with coloradoite and sylvanite. There is no obvious relationship in distribution and occurrence between the sulfide and telluride of lead. Altaite and native gold are very rarely found in contact and yet this pair comprise a major association in ores of the Smuggler and Fourth of July mines near Balarat at the northernmost part of the telluride belt. Here, altaite is veined by gold (Pl. 2b) and both minerals replace older petzite.

All altaite observed in the present study is anhedral, granular, and intergrown with other ore minerals, but imperfect cubic crystals coated with late galena and one large cubic cleavage form ⅝-inch in diameter were

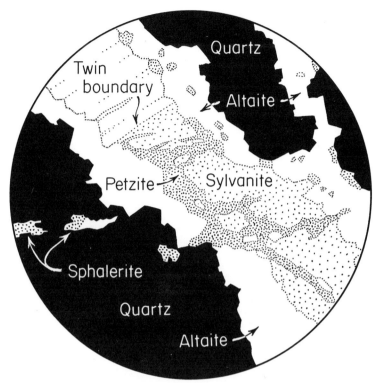

Figure 10. Camera lucida drawing showing sequential replacement of sylvanite by petzite and altaite.

reported from the Red Cloud mine (Genth, 1874) and small cubic crystals were found in the Slide vein (Endlich, 1874; Smith, 1883; Eckel, 1961).

Calaverite

Calaverite has been identified in about 26 percent of the telluride-bearing samples and, where present, averages 16 percent by volume of the metallic assemblages. It is therefore an important source of gold even though it occurs less commonly than sylvanite, petzite, or the native metal.

In view of uncertainties attached to microscopic distinctions of calaverite, krennerite, montbrayite, and untwinned sylvanite, numerous X-ray analyses were performed to ascertain the relative abundance of these minerals. Fifty-two samples, each known to be one of these minerals, were extracted from polished sections and X-rayed. Materials from 24 different mines were represented in this survey which resulted in the following breakdown:

Mineral	Number of Occurrences
Calaverite	28
Untwinned sylvanite	15
Krennerite	9
Montbrayite	0

To the extent that these samples are representative of the ores, it appears that occurrences of calaverite are roughly three times as common as those of krennerite. The results do not in any way reflect the true abundance of sylvanite which was readily identified in many other samples by its pronounced polylamellar twinning, strong bireflection, and diagnostic polarization figure.

In the early mining years, small single crystals of calaverite were found in several mines of Boulder County (Genth, 1877; Endlich, 1878; Smith, 1883; Eckel, 1961). More typically, calaverite is complexly intergrown with the other ore minerals. In polished section, it commonly displays elongate euhedral to subhedral forms against the other tellurides, native tellurium, and younger vein quartz (Pls. 5c and 5d).

Calaverite is very commonly associated with native tellurium, coloradoite, altaite, and melonite and in this type of ore is generally the main source of values. Where intergrown, these minerals produce a variety of puzzling textures that are difficult to describe in any systematic way. In some places the calaverite appears as euhedral crystals embedded in quartz and relatively isolated from the other tellurides and tellurium which occur in nearby vugs. In other specimens, subhedral crystals of calaverite project across cavities in quartz and appear corroded by coloradoite, altaite, and tellurium that occupy those cavities (Pls. 5c and 5d). In the majority of specimens, calaverite occurs entirely in vuggy openings in quartz where it is more intimately intergrown with the other metallic minerals. It may be randomly distributed and mutually intergrown with tellurium, coloradoite, or altaite or concentrated along with the other tellurides toward the borders of tellurium vug fillings. Calaverite and coloradoite are preferentially intergrown in this type of ore, and in many sections they appear together in rims which separate tellurium from surrounding quartz or melonite, or which corrode the outer parts of tellurium crystals in open vugs. Optically oriented inclusions of calaverite in coloradoite are common and in some cases produce a sub-graphic intergrowth (Pl. 5e).

Associations of calaverite and sylvanite are quite rare in the Boulder ores but have been noted in ores of the Buena, Horsefal, and John Jay mines. In the Buena, coarse blades of sylvanite are cut by irregular veinlets of petzite which in places contain minute (.004 to .12 mm) grains of sylvanite, chalcopyrite, and galena (Pl. 10h). Against calaverite, the sylvanite shows markedly stronger bireflectance and pronounced polylamellar twinning. In the Horsefal ores (Pl. 4h), fringes of sylvanite 5 to 15 microns wide appear along contacts of calaverite with later petzite and altaite and in one sample from the John Jay mine similar fringes occur along calaverite-tellurium contacts. Calaverite has been X-rayed in all of these occurrences but the sylvanite could only be identified optically and by tests of microhardness.

Calaverite is associated with native gold and petzite in ores of the Nancy and Emancipation mines. On a gross scale, these minerals display mutual textures (Pl. 5f), but at high magnifications, veinlets of gold can be seen along the grain boundaries in calaverite.

Among the common ore minerals, calaverite does not occur in direct contact with hessite, krennerite, or galena although the pair galena-calaverite has been found in close proximity in the Buena ores described above.

Coloradoite

Coloradoite, the mercury telluride, was discovered as a new mineral in the Boulder veins by Genth (1877), who described impure type material from the Smuggler mine at Balarat and the Keystone and Mountain Lion mines of the Magnolia district. More recent studies have revealed its presence in other mines of the region (Wilkerson, 1939; Lovering and Goddard, 1950; Wahlstrom, 1950), and it is now a well-established species reported in many of the world's important telluride localities. Cuprian coloradoite recently found at Kalgoorlie by Radtke (1963) has not been recognized in the Boulder ores.

Coloradoite compares closely in abundance with altaite, hessite, and calaverite in the specimens studied. The writers have identified it in 64 polished sections, and the list of known occurrences can now be extended to 30 mines within the general area of telluride mineralization (Table 7). Although coloradoite typically occurs in amounts subordinate to other ore minerals, it is the chief metallic constituent of some samples from the Black Rose, Golden Harp, Last Chance, Poorman, and Shirley mines in which it forms massive vug fillings in pyritic horn quartz.

The mercury telluride is hypogene in the Boulder veins and is the primary source of sparse, supergene native mercury occasionally found in oxidized ore. In its varied hypogene occurrences, coloradoite appears consistently as anhedral, granular material intergrown with other metallic minerals.

Coloradoite occurs most abundantly in ore composed of intergrown tellurium, calaverite *or* sylvanite, melonite, and altaite, and appears (1) as rims along the edges of single tellurium crystals or aggregates of tellurium (Pls. 5h and 11c), (2) as irregular patches mutually intergrown with tellurium and sylvanite or calaverite (Pls. 4g and 5d), (3) as anhedral grains either enclosed in single crystals of tellurium (Pl. 7g) or localized along grain boundaries in tellurium aggregates, and (4) as aggregates that show mutual boundaries against tellurium but which conform in whole or part to crystal outlines of the associated gold tellurides (Pl. 10e). The origin of some of these textures is not fully understood, and the age relations of coloradoite are therefore debatable. The problem is one that involves several minerals and will be discussed more fully in later pages.

It is noteworthy that in these tellurium-rich assemblages coloradoite is preferentially intergrown with whichever of the gold tellurides happens to be present (Pls. 2f, 4g and 8b). Descriptions of textures arising from such intergrowths were given under "calaverite" and apply equally well to the similar coloradoite-sylvanite association. As discussed in later pages, this preferential coloradoite-gold telluride association may supplant the strong association of coloradoite and native gold characteristic of the Kalgoorlie deposits (Stillwell, 1931; Markham, 1960). Native gold does vein and replace coloradoite in some ores of the Horsefal, Nancy, and Poorman mines, but the coloradoite is not preferentially intergrown with gold in these few occurrences. Veinlets of native gold are extremely abundant in coloradoite of the Last Chance mine, but here the gold is clearly supergene (Pl. 9d).

In addition to its abundant associations with native tellurium, coloradoite is also common in ores in which sylvanite and petzite predominate. Here it occurs alone or intergrown with petzite in irregular patches, rims, or veinlets replacing older sylvanite (Pl. 6c). In samples from the White Crow, Buena, and Smuggler mines, intergrown coloradoite and petzite form partial pseudomorphs after blades of sylvanite (Pl. 6a). The coloradoite and petzite are mutually intergrown in this type of ore and appear to be contemporaneous.

Coloradoite and hessite are both common minerals, but are antipathetic in occurrence; although found together in a few polished sections, they do not come into direct contact as hypogene minerals. Fine-grained supergene hessite does replace hypogene coloradoite in one sample from the Last Chance mine (Pl. 7e).

In many sections, coloradoite comes into direct contact with the older sulfides and sulfosalts. It is commonly molded against crystals or vug linings of pyrite-marcasite and in places includes corroded relicts of sphalerite, chalcopyrite, tetrahedrite, and galena.

Empressite[4]

Thompson (1949, p. 357) reported "empressite" as sparse disseminations with altaite and pyrite in porphyry from the Red Cloud mine, but subsequent studies by Honea (1964) indicate that the mineral reported is probably stuetzite (Ag_5Te_3). True empressite (AgTe) is the rarest of tellurides in the Boulder deposits, having been found only in one polished section from the Empress mine located in the zone of overlap between the telluride belt and the Boulder County tungsten district (Fig. 1). The type locality for this mineral is the Empress Josephine mine, Kerber Creek district, Saguache County, Colorado, and its occurrence in a mine named the Empress in Boulder County is apparently pure coincidence. The geology of the Empress mine which yielded both gold telluride and tungsten ores has been described by Lovering and Tweto (1953, p. 189-191). In the samples examined, empressite is intergrown with coloradoite and melonite in a small vug about 1 mm long in horn quartz. The empressite is composed of a mosaic of anhedral,

---

[4]Following recent recommendations of Honea (1964), the name empressite is applied here and throughout this report to orthorhombic AgTe, a mineral of very strong bireflectance which gives the X-ray powder pattern of empressite II (Berry and Thompson, 1962). The name stuetzite is applied to hexagonal Ag_5Te_3, a mineral of weak bireflectance optically similar to hessite and which gives the X-ray powder pattern of empressite I (Berry and Thompson, 1962). Diagnostic microscopic properties of these minerals were given in a previous section of this report.

highly bireflectant grains up to 0.3 mm in size. In places, the empressite contains a number of minute graphic voids of unknown origin (Pl. 6d). The coloradoite occurs as anhedral, intergranular inclusions in the empressite. A cluster of fine melonite prisms is in places stretched out across the empressite-coloradoite contacts. Sylvanite and stuetzite occur in nearby vugs scarcely one millimeter away (Pl. 9e), but were not found in direct contact with empressite.

Hessite

Hessite is the leading source of silver in the telluride veins and has accounted for approximately 41 percent of total silver production. Although it averages only 11.3 percent by volume of the metallic assemblages in which it is found, this mineral is present in almost 42 percent of the polished sections examined.

Hessite is quite restrictive in its associations with the other tellurium minerals. It is evidently incompatible with calaverite, krennerite, and native tellurium, as it has never been found in contact with them. Also, it is curious that hessite and coloradoite are not associated, because both are abundant. Aside from these missing combinations, hessite associates freely with the other common ore minerals and its associations with petzite and native gold are particularly strong.

In all available specimens, hessite is intergrown with the other ore minerals and does not display its own crystal form; however, crystals from Boulder County have been described by Palache and others (1944, p. 184). Hessite normally occurs in vugs and drusy seams in quartz where it formed both by replacement of older metallic minerals and by direct precipitation in the available open spaces.

Hessite-sylvanite is one of the most common associations in these ores and gives rise to an interesting variety of textures. Hessite commonly forms veinlets that traverse coarse blades of sylvanite (Pls. 6g and 13d). Many of these veinlets are segmented with areas of petzite or micrographic sylvanite-hessite, or both, that show mutual boundaries against areas of pure hessite. Coarse, tattered inclusions of sylvanite are very commonly found within hessite (Pl. 6h) and micrographic intergrowths of sylvanite in hessite of the type familiar in Kalgoorlie ores (Stillwell, 1931) are known to occur in ten of the Boulder mines. The graphic textures are best developed in ores of the Cash mine in which abundant hessite is peppered with oriented, micrographic inclusions of sylvanite (Pl. 7a). In places, minute inclusions of stuetzite occur as a third phase in these intergrowths (Pl. 7b). Electron photomicrographs of specimens from the Cash vein reveal areas of hessite within the intergrowths which lack the usual inclusions of sylvanite and which are too small to be readily distinguished under the microscope (Pl. 7c).

High-silver ores that the miners called "black tellurium" are comprised of compact intergrowths of hessite and petzite with or without subordinate sylvanite or free gold. Such ore from the Freiberg mine, for example, consists of interlacing seams of hessite-petzite, each up to several millimeters wide, which are woven through fractured, sugary vein quartz. The silver tellurides are here accompanied by minor amounts of tetradymite and gold and traces of sulfides. In extremely rich ore of the Cold Spring-Red Cloud, intergrown hessite and petzite form coatings up to one inch thick on older pyritic vein quartz (Pl. 7d). Black tellurium from the Cash mine consists of hessite-petzite with subordinate amounts of sylvanite. It should be stressed that hessite and petzite display mutual textures in all of these intergrowths and were evidently of contemporaneous origin. Both minerals locally contain veinlets of native gold.

Hessite is very commonly intergrown with sulfides and sulfosalts of the telluride veins and normally replaces these older minerals. Incipient replacement veinlets of hessite commonly occur along sulfide contacts as shown in Plate 12b. In more advanced stages of replacement, relicts of bizarre shapes (Pl. 11f) are all that remain of original sulfosalts replaced by hessite. In some places relicts of sulfides and sulfosalts are so numerous that they impart a finely mottled appearance to hessite in polished section (Pl. 3g). In ore of the Colorado vein, advanced replacement of original pyrargyrite-tetrahedrite by hessite has resulted in complex micrographic intergrowths in which the individual phases are difficult to identify (Pl. 11g).

As a hypogene mineral, hessite is clearly younger than the bulk of common sulfides and sulfosalts. It is distinctly younger than coarse crystals of sylvanite in which it appears as veinlets (Pls. 6g and 13d); as later discussed, experimental studies indicate that these are reaction-replacement veinlets and not exsolution products. Hessite is very commonly veined by hypogene native gold (Pl. 2c). Wherever they are intergrown, petzite and hessite display mutual textures and appear to have been contemporaneous. The origin of hessite-petzite intergrowths and also the hessite-sylvanite micrographic intergrowths are discussed in later pages. Some hessite-petzite intergrowths appear to have been replaced by subhedral crystals of tetradymite (Pl. 11b) but, as explained under "Phase Relations," the complex thermal history and true sequence of events are not altogether apparent from the observed textures.

Most hessite seen in the present study appears to be hypogene and comes from mine levels well below the shallow, poorly developed zone of oxidation. Almost all of it displays twinning that is normally taken as evidence of high-temperature deposition though, as later discussed, there is some question about the reliability of this thermometer. Minor amounts of supergene hessite do occur in partially oxidized ore of the Last Chance mine as extremely fine-grained aggregates replacing hypogene coloradoite (Pl. 7e). This hessite was identified both by electron probe analysis and X-ray methods. It was evidently formed by silver released on oxidation of sylvanite originally enclosed by the coloradoite. This type of hessite and other supergene varieties may have been common close to the surface but very few samples are available to test this possibility.

Krennerite

Krennerite has contributed significantly to gold production in some of the mines, but is the least abundant gold telluride in the Boulder deposits. It has been identified by X-ray methods in only nine samples and very tentatively by optical means in three more. This ore mineral occurs in contact with native tellurium and many of the common tellurides with the notable exceptions of calaverite and hessite. It has never been observed in direct contact with free gold.

Krennerite is intergrown with tellurium and variable quantities of altaite, coloradoite, melonite, and sylvanite in some ores of the John Jay, Melvina, Richmond, and Shirley mines. In this association, krennerite occurs (1) as anhedral to subhedral elongate grains mutually intergrown with tellurium and the tellurides, (2) as elongate, subhedral crystals up to 2 mm long embedded in fine-grained quartz and bordered by irregular rims of tellurium and altaite, and (3) as euhedral to subhedral crystals attached to the walls or extending wall-to-wall in quartz vugs that are otherwise filled by tellurium, altaite, coloradoite, and melonite. In some vugs and drusy seams in quartz, krennerite is involved in a concentric zoning in which the sequence of phases from the walls inward is either quartz-melonite-krennerite-tellurium, quartz-sylvanite-krennerite-tellurium, or simply quartz-krennerite-tellurium. In places, as in Richmond ores, coloradoite is intergrown with and locally veins krennerite in such concentric arrangements.

A different association is represented by ores of the New Rival and Golden Age veins where krennerite is veined and replaced by petzite and altaite. Minute grains of twinned sylvanite commonly appear in such ore and are localized either within the petzite veinlets or along the contacts of coarsely crystalline krennerite with surrounding quartz.

In an unusual specimen from the Alpine Horn mine, small bladed crystals of krennerite have replaced sylvanite and were in turn replaced by late altaite. In polished section, these crystals resemble slender, undeformed needles cutting in all directions across a matrix of deformed sylvanite in which the primary twin lamellae are highly distorted and displaced (Pl. 7f). Electron probe analyses and X-ray powder patterns have confirmed both the krennerite and sylvanite in this intergrowth. This specimen is a significant one; it provides unusually clear evidence for the age relations of krennerite and sylvanite in this particular type of occurrence, and also shows the ease with which euhedral crystals of one telluride can develop in the body of another by replacement.

It is apparent that the age relations of krennerite depend upon the kind of association in which this mineral is found. In the Alpine Horn ore just described, krennerite is definitely younger than sylvanite and is itself replaced by altaite, whereas in the Golden Age and New Rival ores, both sylvanite and petzite are relatively late minerals that vein krennerite. In some intergrowths with native tellurium, krennerite appears contemporaneous with the other metallic minerals, but the textures are rather ambiguous.

Melonite

Melonite in the Boulder veins was first discovered by Hillebrand (1885) in ores of the Forlorn Hope mine. However, the widespread occurrence of melonite was not fully appreciated until the work of Wahlstrom (1950) who identified melonite by X-ray methods in samples from 13 widely scattered mines in Boulder County. Wahlstrom predicted that further study would reveal the presence of melonite in all or most of the telluride deposits. This is certainly borne out by results of the present study in which melonite was identified in 63 polished sections from 31 different mines and prospects. In some unusual samples, melonite makes up as much as 30 percent of the metallic assemblages but is more typically present in amounts less than 5 percent.

Melonite occurs in most places as very fine-grained aggregates conspicuously localized along the borders of native tellurium or early coarse sylvanite, calaverite, and krennerite (Pls. 7h and 8b). This is the type of melonite studied by Wahlstrom (1950, p. 949-950), who describes it as follows:

In polished section, the melonite is an early mineral deposited on the walls of small vugs or druses and stands out in relief against gangue minerals and later ore minerals. The mineral generally appears in thin coatings with a colloform structure, the sinuousities of which reflect in detail the irregularities of the walls of the cavities. The general appearance suggests repeated deposition of thin colloidal layers, but under polarized light the melonite is seen to comprise equigranular or radiating aggregates of tiny, moderately to strongly anisotropic grains.

This description provides an accurate picture of this common type of melonite, but the present authors do not agree with the genetic implications. There is no question that these thin melonite layers give a first impression of early coatings on quartz and this undoubtedly led Wahlstrom to conclude that melonite was the oldest tellurium mineral in the complex sequence. However, there is conclusive evidence in many samples that melonite is actually younger than native tellurium and also the early coarse-grained sylvanite, calaverite, and krennerite. Where seen along the contacts of these minerals with quartz, melonite has replaced the older metallic minerals and is not an early coating on quartz. Several specific examples will serve to illustrate this.

(1) **John Jay Mine.** Hexagonal prisms of tellurium are surrounded by younger vein quartz (Pl. 7g). Melonite and associated coloradoite form thin layers that follow the crystal outlines of tellurium. Viewed at high magnification, these melonite layers give the erroneous impression of having coated the surrounding quartz (Pl. 7h). Note in Plate 7g that the melonite does not appear along contacts of tellurium with older, coarsely crystalline quartz.

(2) **Eclipse Mine.** Coarse blades of tellurium are seen in all stages of replacement by fine-grained melonite which, in places, forms complete pseudomorphs after tellurium (Pl. 8a).

(3) **Ingram Mine.** Veinlets of hessite and petzite transect coarse crystals

of sylvanite that are enclosed by quartz. Melonite occurs only within these veinlets and only where they come into contact with the quartz (Pls. 9b and 13d).

(4) **King Wilhelm Mine.** Melonite occurs only in veinlets of petzite cross-cutting coarse calaverite and only where those veinlets come into contact with roscoelite and quartz (Pl. 8c). The melonite does not appear randomly distributed along quartz-calaverite contacts as would be expected if melonite coated the quartz prior to deposition of the calaverite.

Professor George Tunell has informed the authors of other examples of melonite younger than the gold tellurides as represented by specimens in collections at the University of California at Riverside (Tunell, written commun., 1966). Specimen No. 536 in the Hatfield Goudey Micromount Collection is labelled "Sylvanite and melonite from Magnolia, Colorado," and Tunell describes it as follows:

The specimen is a fragment about 7mm in diameter. It shows a veinlet of drusy quartz (individual quartz crystals each about 0.1 mm in diameter) in a siliceous gangue. Resting on the quartz crystals are numerous euhedral sylvanite crystals (identification by Goudey). The habit appears to be right for sylvanite. The crystals are up to 1.0 mm in length and 0.5 mm in diameter. Some of the sylvanite crystals are very bright and silver white. Other sylvanite crystals are coated by aggregates of extremely fine melonite crystals and some sylvanite crystals are embayed and partly replaced by the melonite aggregates.

Several other specimens, from the 2100, 2200, and 2300 levels of the Cresson mine at Cripple Creek in the George Tunell Collection, show pseudomorphs of melonite and native gold after calaverite.

By stressing examples of replacement by melonite in the Boulder County ores, the writers do not intend to imply that all of the melonite formed in this way. In some cases the melonite grew in open vugs but here tended to form larger single crystals (up to 1 mm long) rather than the fine-grained aggregates with "colloform structure" as described by Wahlstrom (1950). Such coarse melonite occurs in ores of the Alpine Horn, Cash, Cold Spring-Red Cloud, Horsefal, and Winona mines (Pls. 8e and 8f). Ore minerals definitely later than the melonite (e.g., hessite, coloradoite, altaite) fill in around the coarse melonite crystals (Pls. 8e and 8f) and in some cases corrode them slightly (Pl. 8f). In one specimen from the John Jay mine (Pl. 8g) small vugs in quartz were first lined by an intergrowth of quartz and fairly coarse melonite and then filled centrally by coloradoite. A thin film of quartz formed over the quartz-melonite intergrowth before introduction of the coloradoite.

In summary, the writers believe that melonite formed at an intermediate stage in the telluride sequence after native tellurium and the early gold tellurides, but during and largely before deposition of altaite, coloradoite, hessite, petzite, and native gold. Where nickel-bearing solutions encountered existing tellurides or native tellurium, melonite developed as fine-grained aggregates replacing the older ore minerals along free surfaces or along their contacts with gangue. Where these same solutions entered open cavities, melonite grew as fewer crystals that generally attained much larger dimensions.

Nagyagite

Rare nagyagite occurs in several veins of the Jamestown and Gold Hill districts where it is of little more than mineralogical interest. In these few occurrences, it is consistently associated with the silver tellurides, hessite, or petzite.

At the King Wilhelm mine, small platey crystals of nagyagite up to .1 mm in diameter are embedded in petzite; in polished section these plates appear bent and many are joined at high angles forming V- and T-shaped clusters (Pl. 9a). Many of the plates are attached to quartz of surrounding vug walls or to early crusts of sulfides and sulfantimonides on the walls.

A trace of nagyagite occurs in ore of the Black Rose mine where some 13 metallic minerals are intimately intergrown. Minute laths of nagyagite occur only within hessite in this complex assemblage.

In one sample of Little Johnny ore, a network of tiny warped plates of nagyagite (.01 to .1 mm in diameter) occur in altaite, calaverite, and petzite, but are most abundant in petzite. An almost identical occurrence was noted in a sample from the 350-foot level of the Buena mine in which nagyagite crystals measuring up to .52 mm are enclosed in both petzite and calaverite but are concentrated in the petzite.

Nagyagite is a sparse ingredient of ore from the Nancy mine. Here, it occurs in contact with hessite, petzite, coloradoite, and gold in fillings of vugs that are lined by pyrite-marcasite. Petzite penetrates the nagyagite along cleavage (Pl. 8m). Native gold veins hessite and petzite in this ore and in places is localized along nagyagite-coloradoite contacts.

The age relations and origin of nagyagite are not fully understood. Its close and consistent association with hessite and petzite suggests a timing and origin related in some way to the formation of those minerals.

Petzite

Economically, petzite is one of the most important ore minerals in the Boulder County deposits; among the tellurides it ranks second to sylvanite as a source of gold and second to hessite as a source of silver. Petzite occurs in about 45 percent of the samples investigated in amounts averaging 12.9 percent of the metallic assemblages.

This mineral never displays good crystal forms in the Boulder ores, but rather occurs as dense, anhedral aggregates intergrown with other ore minerals. It occurs in contact with all of the other abundant gold-silver tellurides and is most commonly associated with sylvanite and hessite. It has never been seen in contact with native tellurium but is commonly veined and replaced by native gold.

A significant proportion of petzite in the veins occurs as veinlets, rims, and irregular masses replacing older, coarsely crystalline sylvanite, calaverite, or krennerite. Where present as veinlets, the petzite is normally confined to the body of the gold telluride host (Pls. 9b and 11a), but in some cases extends out into adjacent vein sulfides or sulfosalts (Pl. 10b). Petzite that has replaced sylvanite is commonly intergrown with late hessite, or graphic sylvanite-hessite. Petzite that has replaced calaverite locally contains minute inclusions of chalcopyrite or sylvanite (Pl. 10h), or in some sections is veined and replaced by native gold, but where calaverite is the host, hessite is not present. Petzite that has replaced krennerite also contains tiny grains of sylvanite, but in this association is never seen in contact with hessite or gold. Coloradoite may accompany petzite in all of the settings described above; it appears either contemporaneous with or somewhat later than the petzite (Pl. 5g).

The formation of petzite replacement rims is best illustrated by ores of the Smuggler mine in which coarse blades of sylvanite are bordered by rims of younger petzite which, in turn, are replaced by outer rims of late altaite (Fig. 10). Coloradoite is mutually intergrown with petzite in some rims of this kind (Pl. 12d), but in other places (Pl. 6b) forms distinct rims of its own that encroach upon petzite. In the advanced stages of such replacements, partial or complete pseudomorphs of intergrown petzite and coloradoite in places preserve the prismatic forms of original calaverite or sylvanite crystals (Pl. 6a).

Compact vug fillings of petzite and intergrown petzite-hessite are very common. In ores of the Osceola-Interocean workings, for example, intergrown petzite and hessite occur in vugs lined by older pyrite and melonite (Pl. 8d) and in some samples from the Nancy mine almost pure petzite seals vugs that contain early linings of pyrite-marcasite (Pl. 4c). Petzite of the Nancy mine contains some scattered corroded relicts of pyrargyrite (Pl. 11f) and in a few places is intergrown with rare nagyagite (Pl. 8h). The richest silver ore seen by the authors is composed of intergrown petzite and hessite which form thick coatings on pyritic vein quartz of the Cold Spring-Red Cloud mine (Pl. 7d). Such high-grade silver telluride ore is rare and, as previously mentioned, is called "black tellurium" by the miners.

Petzite veins the youngest of the common sulfides, chalcopyrite and galena, (Pl. 3f) as well as the early gold and gold-silver tellurides, sylvanite, calaverite, and krennerite (Pls. 9b, 10h and 11a). In a few samples, very small grains of sylvanite appear along grain boundaries in petzite (Pl. 10d) or along petzite-hessite contacts. Petzite is rather commonly veined and replaced by hypogene native gold and is more rarely veined by altaite (Pl. 5b).

The unusual relationship of petzite and hessite should be stressed. The two minerals are closely associated in many ores and invariably display mutual textures; the intimate intergrowths give the impression of having formed by separation of the two minerals from some original homogeneous phase. This possibility is further examined in later pages in connection with phase relations that involve petzite.

Rickardite
Traces of rickardite occur as a supergene alteration product of either

sylvanite or calaverite in ores of the Buena, Horsefal, Last Chance, Poorman, and Potato Patch mines. This mineral occurs in such small amounts or is so contaminated that the writers have never been able to establish its identity by X-ray methods. However, on the basis of its distinctive colors and strong reflection pleochroism (red-purple to gray), extreme anisotropism, vivid polarization colors (orange to blue to gray), and significant reaction to KCN (slowly turns gray-black), there is little doubt that this mineral is rickardite.

In all observed occurrences, rickardite is closely associated with supergene free gold and tellurium oxide, either in intimate intergrowth or at least in close proximity. In some samples, rickardite appears as an incipient alteration product along the edges of calaverite or sylvanite grains (Pl. 9c), and, in oxidized ore from the Potato Patch, intergrowths of gold and rickardite occur in tellurite replacing native tellurium and apparently mark the sites of original gold telluride grains (Pl. 12e).

In partially oxidized ore of the Last Chance mine, gold-rickardite intergrowths of the type described at Kalgoorlie by Stillwell (1931) apparently formed by replacement of original sylvanite. Gold forms a network of filaments and veinlets within the rickardite and is in places the main constituent of the intergrowths. The intergrowths form curiously shaped inclusions in coloradoite that are largely unaltered (Pl. 9d). Some features point to sylvanite as the original telluride: (1) in places, the gold-rickardite intergrowths have a lamellar structure reminiscent of twinning in sylvanite and (2) the formation of supergene hessite replacing coloradoite near these intergrowths suggests an original silver mineral.

Stuetzite

A rare silver telluride optically similar to hessite occurs in small quantities in a few samples from the Black Rose, Cash, Empress, and Smuggler mines. The microscopic properties and X-ray powder pattern of this mineral match those of stuetzite (Ag_5Te_3) as redefined by Honea (1964).

In ores of the Smuggler, stuetzite is intergrown with sylvanite, tellurium, and coloradoite, and all four minerals form compact fillings of numerous small vugs in fine-grained quartz. In most places, the stuetzite displays mutual contacts with the other ore minerals, but in a few vugs it veins sylvanite and also forms rims surrounding elongate crystals of both tellurium and sylvanite (Pl. 9g).

In one sample from the Black Rose mine, stuetzite is one of six tellurides found replacing a complex assemblage of sulfides and sulfosalts. The sylvanite and stuetzite are in mutual contact, and are veined and replaced by hessite (Pl. 9f).

Stuetzite is associated with micrographic sylvanite-hessite in one sample from the Cash mine. It forms irregular patches 10 to 40 microns across at the edge of vugs occupied by the curious sylvanite-hessite intergrowths previously described. Within these intergrowths, minute grains of stuetzite appear welded to the inclusions of sylvanite (Pl. 7b).

Small areas of stuetzite up to .1 x .4 mm in size appear in a polished section of ore from the Empress mine. Here, the stuetzite occurs along with coloradoite as anhedral patches along grain boundaries in sylvanite and also appears in mutual intergrowths with coloradoite-sylvanite (Pl. 9e). In places, minute fibers and prisms of melonite appear within the stuetzite, sylvanite, and coloradoite.

As shown by Honea (1964), the "empressite" identified by Thompson (1949) in a sample from the Red Cloud mine is very probably the mineral stuetzite (Ag_5Te_3). The mineral described by Thompson occurs as disseminations with altaite and pyrite in porphyry.

Sylvanite

Sylvanite was identified in ores of the Grandview and Red Cloud mines shortly after the initial discoveries of telluride ores in the Gold Hill district. As pointed out by Eckel (1961, p. 316), these were the first known occurrences of sylvanite in the United States and were described almost simultaneously by Genth (1874) and Silliman (1874a and b). As mining progressed, this mineral proved to be the most widespread and abundant telluride and is probably present in variable amounts in all of the telluride veins. It easily ranks as the most important telluride economically, as it has accounted for approximately 22 percent by weight of gold and of silver extracted from the hypogene ores. In the present study, sylvanite was identified in 93 polished sections representing some 38 mines and prospects throughout the county and, where present, it averages 20.4 percent by volume of the metallic assemblages.

In the early mining days, fairly pure single crystals of sylvanite were found in some of the mines, and the older literature contains many descriptions and some chemical analyses of these rare finds (Genth, 1874; Silliman, 1874a and b; Clark, 1877; Endlich, 1878). High-grade ore from the Walker-Clarke stope of the Buena mine contains small, well-formed crystals of sylvanite and of tellurium scattered through cavities in a porous intergrowth of quartz, roscoelite, and fluorite. The sylvanite crystals are 0.4 to 0.6 mm in diameter and up to 4 mm long. These appear to be pure in hand specimens, but contain microscopic inclusions of altaite, coloradoite, and other metallic minerals. One of these crystals from the Buena was used by Tunell (1941) in his studies of the atomic structure of sylvanite.

Other occurrences of well-formed sylvanite crystals were noted in ores of the Poorman, Ingram, Golden Bell, and Franklin mines. Many samples from the Ingram contain very thin, elongate blades of sylvanite which measure on the average .1 x 5 x 25 mm. These crystals lie in drusy seams lined by vein quartz and are arranged with their broad faces roughly parallel to the vein structure (Pls. 9h and 10a). The sylvanite appears "plastered" against the older quartz and in many places is perforated by euhedral quartz crystals that are about 1 mm long. Younger and finer-grained quartz crystals (.3 to .6 mm long) locally coat and envelop these fragile sylvanite crystals.

Crystals of similar habit but somewhat larger dimensions (.7 x 10 x 30 mm) were observed in one sample from the Poorman mine and these are in places coated by late calcite. In other samples from the Poorman, and also in ores of the Franklin and Golden Bell, fairly pure crystals of sylvanite grew in open vugs in quartz and were cemented by later fillings of ankerite and calcite. These crystals are typically 1 mm long and have a skeletal character produced by deep grooves and recesses running parallel to the long dimensions. In polished section, this sylvanite shows sub-graphic forms and round holes appear within the crystals that reflect the presence of interior tubular openings that also run the length of the crystals.

In more typical ores, sylvanite is intimately intergrown with the other ore minerals and pure samples are difficult to obtain. In one place or another, sylvanite occurs in contact with all of the other tellurides with the exceptions of sparse tetradymite and nagyagite. Sylvanite is incompatible with native gold as evidenced by the fact that these common minerals are never seen in direct contact. Sylvanite is most abundant either in association with hessite-petzite or with native tellurium, and these distinctly different assemblages do not intergrade (i.e., neither hessite nor petzite occur in association with free tellurium). In either type of assemblage, sylvanite may be accompanied by altaite, coloradoite, and melonite, although these minerals are noticeably more abundant where sylvanite is associated with native tellurium.

Most sylvanite is early in its associations with hessite, petzite, coloradoite, and altaite, and hence is subject to replacement by these younger minerals which appear in veinlets, rims, and pseudomorphs. In many samples, early bladed sylvanite is veined by petzite, hessite, coloradoite, and micrographic sylvanite-hessite (Pls. 9b and 13d). These veinlets consist either of a single phase or are segmented and contain two or more of the late phases mentioned. Rarely, such veinlets extend through sylvanite and continue out into adjacent sulfide or sulfosalt minerals (Pl. 10b). Rim replacement of sylvanite is well illustrated by ores of the Smuggler mine where, in some samples, slender blades of sylvanite up to 5 mm long occupy a common network of fractures in horn quartz (Pl. 12c). In polished section, these crystals are completely rimmed by younger petzite which is in turn replaced by outer rims of late altaite (Fig. 10). In other samples from the Smuggler, these rims have completely replaced the sylvanite whose bladed form is then preserved by intergrown altaite-petzite (Pl. 10c). Similar pseudomorphs are seen in ores of the Logan mine where the replacing minerals are coloradoite and petzite (Pl. 6a).

Minor amounts of sylvanite formed late in these sylvanite-hessite-petzite ores. For example, the micrographic inclusions of sylvanite in hessite are younger than coarse bladed sylvanite which is commonly veined by the micrographic intergrowth. The orientation of sylvanite grains within the intergrowths bears no obvious relationship to the orientation of older sylvanite in the adjacent walls. Also, traces of sylvanite in the Smuggler and

Black Rose ores may be late where it occurs as minute grains localized along hessite-petzite contacts or along grain boundaries in petzite (Pl. 10d).

The textures observed in ores where sylvanite is associated with native tellurium are varied in detail, and in many respects are similar to those described for calaverite in its corresponding associations with native tellurium. The sylvanite occurs either as euhedral to subhedral crystals scattered through and enclosed by younger vein quartz, or as subhedral to anhedral grains of variable size intergrown with other metallic minerals and confined to cavities within older vein quartz. The first case is exemplified by some ore of the Alpine Horn mine in which small, .1 x .5 mm, crystals of sylvanite are more or less disseminated through quartz which carries intergrown coloradoite-tellurium only within vugs. Where sylvanite locally projects into these vugs, it gives the impression of having been corroded by the coloradoite and tellurium. In the second case, where sylvanite is also confined to vugs in older quartz, it appears (1) as subhedral, elongate grains up to .5 mm long attached to the vug walls and projecting inward across fillings of anhedral coloradoite and tellurium (Pl. 10e), (2) as tiny elongate to irregular grains 3 to 10 microns long, concentrated in bands of coloradoite that border grains or aggregates of native tellurium (Pls. 2f, 8b and 10f), and (3) as anhedral grains irregularly distributed and mutually intergrown with coloradoite and native tellurium. Where present, melonite occurs along the borders of these sylvanite-tellurium-coloradoite intergrowths (Pl. 8b) and altaite appears as a later mineral replacing the sylvanite, coloradoite, and tellurium (Pls. 4g and 8b).

The textures and associations of sylvanite in this type of ore are so similar to those of calaverite in its associations with native tellurium that these gold tellurides appear to proxy for one another in these tellurium-rich assemblages. Sylvanite and calaverite are themselves very rarely intergrown in the presence of free tellurium. An example of this rare combination appears in a sample from the John Jay mine where narrow bands of twinned sylvanite up to 20 microns wide separate calaverite from direct contact with tellurium.

The associations sylvanite-krennerite-petzite and sylvanite-calaverite-petzite have been observed in a few samples, but these are comparatively uncommon. The minute grains of sylvanite that occur in petzite veinlets transecting coarsely crystalline krennerite or calaverite (Pl. 10h) were previously described. Sylvanite also occurs in narrow bands 5 to 15 microns wide separating calaverite from younger altaite and petzite in one sample from the Horsefal vein (Pl. 4h).

Very slight warping and microscopic offsets of primary twin lamellae in sylvanite appear in some samples, but most of the sylvanite suffered little deformation during and after its initial deposition. Samples from the Alpine Horn mine, as shown in Plate 7f, display an unusual degree of deformation. The occurrence of late krennerite crystals that replace the highly deformed sylvanite in this ore is also quite unusual.

The age and textural relations of sylvanite with respect to the common vein sulfides and sulfosalts are clear-cut. Pyrite and pyrite-marcasite intergrowths commonly border or are enclosed by younger sylvanite in vugs and show little sign of replacement. However, sphalerite, tetrahedrite, galena, and chalcopyrite are generally veined and extensively replaced by the younger telluride. In ores of the White Crow and Bondholder, traces of unusually late chalcopyrite appear as veinlets along grain boundaries in sylvanite (Pl. 3e). Also, in a few specimens coarsely crystalline tetrahedrite and sylvanite are intergrown in such a way as to suggest contemporaneity (Pl. 11a). Contrary to these few exceptions, the evidence in the great majority of polished sections indicates that the main stage of sulfide deposition had come to a close before sylvanite was precipitated.

Tetradymite

As a group, the tellurides of bismuth are poorly represented in the Boulder County deposits, but minor amounts of tetradymite do occur in some of the veins. This mineral was first recognized in ores of the Red Cloud mine by Genth (1874) and later found by Thompson (1949) in a compact mass of tetradymite, petzite, and pyrite from the same property. In the present study, tetradymite has been identified in ores of the Emancipation, Freiberg, Red Cloud, and Stanley mines and in each case the identification was verified by X-ray methods.

In the Red Cloud and Freiberg occurrences, tetradymite appears in a dense, silver-rich intergrowth of hessite, petzite, and pyrite. The tetradymite forms tattered laths and plates up to 2 mm long that extend across the hessite-petzite contacts (Pl. 11b). In places, fine crystals of tetradymite with minute inclusions of chalcopyrite are localized along contacts of hessite and petzite with late crystals of aikinite ($PbCuBiS_3$) that have developed along free surfaces of the silver telluride intergrowths (Pl. 11d). The tetradymite appears to have formed as a reaction rim during replacement of hessite-petzite by aikinite but, as later discussed, the origin of these rims may be somewhat more complex than immediately suggested by the textures.

An entirely different association of tetradymite is represented by ores of the Emancipation mine in which the mineral is intergrown with tellurium, melonite, coloradoite, and calaverite. The tetradymite veins (Pl. 11c) and marginally replaces tellurium, and is itself replaced by the late sponge-variety of native tellurium (Pl. 3b).

In several samples from the dump of the Stanley mine, tetradymite is associated with coarse pyrite, bismuthinite, and free gold, but this occurrence probably represents an unusual phase of the older pyritic gold mineralization and was not formed during the main stage of telluride deposition. ·

SULFOSALTS

Aikinite

Small amounts of aikinite were identified by X-ray methods in ores of the Cold Spring-Red Cloud mine. This mineral occurs as inconspicuous, anhedral, anisotropic grains along free surfaces of a rich hessite-petzite-pyrite intergrowth. In polished section, the aikinite is separated from the silver tellurides by a narrow persistent layer of tetradymite and sparse chalcopyrite (Pl. 11d).

Jamesonite

Traces of jamesonite occur in one sample from the Poorman mine and there form narrow fibrous layers selectively localized along tetrahedrite-sphalerite contacts (Pl. 11e).

Kobellite

Eckel (1961, p. 200) reviews previous, but unconfirmed reports of kobellite in the Boulder ores and points out the questionable validity of this mineral. The early record carries reports of kobellite in the Magnolia district (Dana, 1892) and in the Moscow mine at Sugarloaf (Smith, 1883). No kobellite has been found in the present study.

Pyrargyrite

Minor amounts of pyrargyrite are intergrown with tellurides in samples from the Colorado, Gray Copper, Ingram, and Nancy mines, but nowhere is this sulfosalt abundant in the telluride stage of mineralization. Where it is directly associated with tellurides, pyrargyrite is preferentially intergrown with and replaced by petzite or hessite. Most commonly, the pyrargyrite occurs as small inclusions (up to .4 mm in size) in hessite and shows bizarre, irregular shapes (Pl. 11f). Pyrargyrite occurs most consistently in ore of the Colorado vein where it appears along with tetrahedrite as abundant, minute inclusions in hessite (Pl. 11g). At first glance, the pyrargyrite was mistaken for stuetzite which has a similar blue-white color and is also enclosed by hessite, but distinct red internal reflections are seen in the pyrargyrite. The identification was finally confirmed by X-ray analysis of a single large inclusion found in ore from the Nancy mine.

In ores of the Gray Copper mine, the pyrargyrite inclusions are molded against pyrite, appear to replace sphalerite, are mostly enclosed by hessite, and are locally veined by gold. Pyrargyrite from the Nancy mine shows a few lamellar twins and is replaced by both nagyagite and hessite.

Tetrahedrite

Tetrahedrite is the only sulfosalt that occurs with any persistence in the telluride veins, but even it appears in very small quantities. It is present in 28 percent of the telluride-bearing polished sections examined but, on the average, makes up only about 3 percent of the assemblages in which it occurs.

In most samples, tetrahedrite is accompanied by one or more of the common sulfides—chalcopyrite, galena, and sphalerite. It normally veins sphalerite (Pl. 11e), and is in turn veined and replaced by both galena and chalcopyrite (Pl. 3d). Traces of tetrahedrite occur as relicts in the younger tellurides and native metals and in numerous sections this mineral is veined by sylvanite, hessite, petzite, and gold (Pl. 10b). Where tetrahedrite is replaced by hessite, the contacts between the two minerals are very sinuous and advanced replacement leads to a micrographic intergrowth of tetrahedrite in hessite. Complex intergrowths of tetrahedrite and pyrargyrite in ores of the Colorado mine, just described under pyrargyrite (Pl. 11g), illustrate the intricate textures arising from replacement of the sulfosalts.

Most of the tetrahedrite in the telluride deposits is anhedral and intergrown with the other ore minerals, but well-formed, skeletal crystals were found in one sample from the Poorman mine. These occur with crystals of sylvanite of comparable size in drusy openings in vein quartz. Some of the sylvanite and tetrahedrite crystals are intergrown and in polished section display mutual textures suggestive of contemporaneous deposition (Pl. 11a). Both minerals are veined by petzite (Pl. 10b) and in places are overgrown by coarse, clear calcite that formed as a late vug mineral.

Five pure samples of tetrahedrite from different mines were X-rayed and tested microchemically. All of these gave strong reactions for copper, iron, and antimony but negative tests for silver, lead, and arsenic. Argentian tetrahedrite, an economically important mineral in the lead-silver veins, apparently does not occur in association with the telluride minerals.

Other Sulfosalts

A variety of other sulfosalts including argentian tetrahedrite, miargyrite, polybasite, proustite, pyrargyrite, and stephanite occur in the lead-silver deposits. As a rule, these ores occur in veins separate from the telluride deposits, but in some cases both kinds of ore occur in the same veins and even the same hand specimens. Even where they are this closely associated, the telluride and lead-silver minerals occur in distinctly separate assemblages and are housed in different generations of vein quartz. In samples from the Croesus, Grandview, and Ingram, these lead-silver sulfosalts are contained by milky quartz older than the horn quartz which is associated with the tellurides. The textures and associations of the lead-silver ores have been described by several authors (Lovering and Goddard, 1950; Lovering and Tweto, 1953, p. 39-41) and were not reinvestigated in the present study.

The properties and occurrence of an unknown sulfosalt intergrown with tellurides in the Croesus, King Wilhelm, and Little Johnny mines are described in a later section on unidentified minerals.

HALIDES

Calomel

All available samples of telluride ore were examined in ultraviolet light in a search for fluorescent minerals that might otherwise be overlooked. In the process, minute gray flecks and films of calomel were detected in cavities and on surfaces of partially oxidized ore from the Poorman mine and were confirmed by X-ray analysis. Globules of native mercury accompany the calomel in cavities and the sample contains abundant unoxidized coloradoite. This mineral is extremely rare in the Boulder ores.

Fluorite

Fluorite is common in some of the telluride deposits, but only as a subordinate gangue mineral. It has been reported in telluride veins as far south as those of the Magnolia district (Wilkerson, 1939), but is common principally in a few mines of the Jamestown district in or close to ground strongly mineralized during the main stage of fluorite deposition that preceded the telluride ores.

Fluorite occurs in telluride ore either as breccia fragments of an older stage cemented by telluride vein matter or as a mineral formed as part of the telluride mineralization itself. These relationships of early and late fluorite are well illustrated in the Buena mine where, according to Lovering and Goddard (1950), purple fluorite is common in the telluride veins where they intersect older, north-south fluorite-bearing veins. In polished sections of Buena ores, fluorite occurs (1) as angular purple fragments cemented by a tan-colored, opal-bearing gouge that is itself replaced by the tellurides and (2) as well-formed purple cubes up to 1 mm in size which are intergrown with roscoelite and in places coated by the tellurides. In one Buena sample, a third generation of fluorite is represented by pale blue crystals lodged in cavities within the roscoelite-telluride intergrowths. The blue fluorite appears to have formed after the ore minerals.

Minor amounts of fluorite also occur in John Jay ores, and here are seen cementing fragments of early vein quartz and pyrite. The fluorite is cut by veinlets that contain younger quartz, barite, and the tellurides. In one sample from this mine, a single angular fragment of purple fluorite is cemented by telluride-bearing, pyritic horn quartz.

Veinlets found in stope fill from the 200-foot level of the Melvina mine show a definite sequence in which purple fluorite coats quartz-adularia and is itself coated by younger ankerite, calcite, and gypsum in that order.

Most fluorite of the telluride veins appears a uniform deep violet to dark purple in hand specimen, but, in thin section, small .02 mm clots with a black-purple color are visible and these probably center around minute pitchblende inclusions as suggested by Goddard (1946).

Silver Halides

Minor amounts of cerargyrite, "embolite," and iodyrite were reported by early workers as rare fragments and coatings in oxidized telluride ores of the Gold Hill district (Endlich, 1874, 1878; Smith, 1883). Unfortunately, the present authors have very few samples of the near-surface ores removed 90 years ago, and hence have been unable to confirm these early reports.

CARBONATES

A variety of carbonates, including ankerite-dolomite, calcite, siderite, and rhodochrosite occur in the telluride ores and, although they generally appear in amounts subordinate to vein quartz, these minerals are locally abundant and comprise a significant part of the gangue. Although some ankerite-dolomite is older than the ellurides, the main precipitation of carbonates followed deposition of the or minerals. Most of the carbonate in the veins occurs (1) as late fillings of vugs in telluride-bearing quartz, (2) as central fillings of banded veins separating symmetrical and older bands of ore-bearing horn quartz, and (3) as veinlets of pure carbonate which crosscut older telluride seams or transect telluride-bearing quartz. The latter veinlets are normally thin and inconspicuous, but in ores of the Emancipation, Colorado, and Ingram mines, attain thicknesses up to one inch. In these different occurrences, the vein carbonates may be coated by a number of younger minerals including gypsum, opal, and clays.

Ankerite-Dolomite

Members of the ankerite-dolomite series are most abundant. For want of extensive X-ray data, these minerals are grouped as "ankerite-dolomite," but numerous tests of specific gravity suggest that their compositions span the entire range from ankerite to dolomite. Microchemical tests did not reveal the presence of manganese in carbonates of this group. In many specimens, coarse crystals of ankerite-dolomite show color zoning in which darker brown ankeritic cores grade outward into white dolomite. Most ankerite-dolomite formed after free gold and the other ore minerals, but at least minor amounts are older as evidenced by one specimen from the Richmond mine in which the crystals of ankerite are enclosed and corroded by native tellurium and associated tellurides (Pl. 11h). Ankerite-dolomite is inter-grown with native gold in many samples from the Emancipation, Evening Star, Golden Harp, New Era, and Slide mines, and in these the carbonate fills in around leaves, wires, or irregular masses of gold. We have sampled this type of ore at a depth of 1100 feet in the Slide mine far below the zone of oxidation.

Calcite

Calcite is locally abundant and is younger than ankerite-dolomite. In banded ore of the Colorado vein, calcite forms a central filling of veins with outer crusts of ankerite-dolomite (Pl. 13c). Age relations are also clear in John Jay ores in which veinlets of calcite cut across veinlets of ankerite-dolomite, and, where the two carbonates occur in vugs, calcite coats ankerite-dolomite. In ores of the Franklin and Poorman, clear calcite has overgrown and in places veins crystals of sylvanite lodged in cavities in sugary vein quartz. The calcite described is probably hypogene, but supergene calcite also has been observed in partially oxidized ores from the Poorman mine. Here, the calcite occurs as minute, rusty-brown rhombohedra lining polygonal cavities in vein quartz that were apparently formed by leaching of original fragments of gangue or country rock. This calcite is associated with stains and coatings of limonite and tellurite.

Siderite

Siderite occurs as an early wall rock alteration product but is rare as a vein mineral. In one sample from the third level of the Cash mine (210 feet vertically below the surface) siderite forms fine-grained coatings on petzite and associated sulfides. Siderite also coats the ore minerals in one sample from the Poorman mine. In ores of the Cold Spring-Red Cloud, siderite occurs along with gold and limonite in cavities within hessite-petzite and appears to be supergene.

Rhodochrosite

Rhodochrosite is rare, but has been reported along with ankerite in vein-lets crosscutting telluride-bearing horn quartz of the Ingram mine (Lovering and Goddard, 1950). The present authors have seen one sample from the Ingram in which coarse, pink rhodochrosite coats pyritic horn quartz and is itself coated by younger layers of ankerite-dolomite. Unusual brownish green clots of muscovite (?) up to 1 mm in diameter occur on the surface of the horn quartz and are overgrown by the younger carbonates.

SULFATES

Barite

Barite is locally a minor gangue mineral, and has been identified in ores of the Alpine Horn, Colorado, Herald, Ingram, John Jay, Melvina, and Shirley mines. This mineral was X-rayed in only one sample, but was readily identifiable in thin sections and its presence was tested in many hand samples by barium flame tests.

Barite is an early mineral in the telluride veins and most of it appears to have formed at about the time when the iron sulfides were precipitated. In general, it occurs as subhedral, bladed crystals up to 5 mm long encased in vein quartz.

Typical age relations of barite are revealed in a symmetrically banded vein of the Colorado mine in which fan-like aggregates of barite crystals are intergrown with quartz and form early crusts along the borders of the vein (Pl. 13c). Silicified granite bordering this vein contains adularia which is presumably older than barite. The quartz-barite crusts are succeeded by successively younger layers of pyrite, quartz-pyrite, ankerite, and calcite toward the center of the vein. The tellurides and native gold locally replace all of the vein minerals older than ankerite, but are concentrated along the contact of the pyrite and quartz-pyrite layers.

Tabular crystals of barite several millimeters long occur in tellurium-rich ores of the John Jay mine. The barite here is attached to older vein quartz and colloform aggregates of marcasite, and all three minerals are enveloped by younger fine-grained quartz. Native tellurium and associated tellurides in places corrode the older barite crystals.

At least traces of barite formed after deposition of the tellurides. This is evidenced by ores of the Shirley mine of the Eldora district in which microscopic blades of barite occur only in cavities in an older intergrowth of quartz, roscoelite, adularia, and the tellurides.

Gypsum

Gypsum is common as a supergene mineral in near-surface ores. Fine-grained gypsum commonly grouts fractures crosscutting telluride ore or appears as delicate fibers perched in open seams and cavities in the ores. The mineral evidently formed wherever calcium in the groundwaters encountered sufficient concentrations of sulfate derived from oxidation of pyrite-marcasite.

Rare selenite occurs in abundance 900 feet below the surface in the tenth level of the Ingram mine. The coarsely crystalline gypsum forms a late central filling of a banded vein in which the sequence is (1) coarse, bull quartz, (2) pyritic horn quartz, (3) ankerite, and (4) selenite. Gypsum may be a late hypogene mineral in this unusually deep occurrence.

J. E. Byron provided the authors with a sample labeled "Magnolite, Keystone mine, Magnolia" which fitted perfectly with Genth's original description of that mineral from the same mine (1877). This mineral was originally thought to be a tellurate of mercury but, according to Eckel (1961), has long been discredited. The "magnolite" occurs as radiating groups of thin white crystals (.05 $\times$ .3 mm) perched on dark brown limonite coating tellurides. This "magnolite" gives an excellent X-ray powder pattern for gypsum.

Jarosite

Supergene jarosite is extremely common in partially oxidized ore, and is the mineral responsible for the characteristically yellow stains and coatings seen on the mine dumps. Oxidation of abundant fine-grained pyrite and marcasite in the presence of hydromica and sericite in the altered wall rocks of the telluride veins evidently favored the development of jarosite. Jarosite comprises the yellow stains that were sought by the miners as a guide to pyritic horn quartz that normally carries the tellurides. In the present search for tellurides in the mine dumps, the best finds have been in samples stained with jarosite.

Jarosite is normally associated with limonite (goethite) and in most specimens these two minerals display an interesting zonal arrangement. Where they impregnate both horn quartz and adjacent wall rock, jarosite predominates at the site of partially decomposed iron sulfides and grades outward into limonite. Where these minerals form transported linings of vugs or other openings, jarosite generally forms a distinct yellow inner lining as though it were the later mineral. As stains along fractures, jarosite and limonite are commonly very irregularly distributed but still show distinct segregation. Chemical reasons for these curious jarosite-limonite relations will be offered in later pages.

Several jarosite samples from different mines were X-rayed and found to contain minor gypsum as a contaminant. X-ray fluorescence analyses revealed traces of silver and lead in all of the samples but the form of occurrence of these elements is unknown.

TELLURITES AND TELLURATES

The early literature contains reports of tellurites and tellurates in Boulder County, but none of these has been substantiated by X-ray analysis.

Genth (1877) tentatively identified a yellow mineral on calaverite from the Red Cloud mine as montanite $[(BiO)_2(TeO_4) \cdot (2H_2O)]$, and in view of the fact that this is one of the few mines where the bismuth telluride, tetradymite, is moderately abundant, the finding of montanite here seems quite reasonable.

Genth (1877) also described a straw- to lemon-yellow mineral associated with native tellurium of the Keystone mine at Magnolia for which he suggested the formula $FeTeO_4$ and the name ferrotellurite. As noted by Frondel and Pough (1944, p. 217), Genth later stated in a personal communication to E. S. Dana (1892) that the yellow material might have been tellurite colored by some salt of iron. Noting the lack of definitive characters for this mineral, Frondel and Pough (1944) recommended that it be dropped from the record. It is relevant that all of the tellurite seen in the present study is colored a pale to bright yellow, presumably by traces of iron. No iron tellurites or tellurates were detected in X-ray analyses of such materials.

"Magnolite," a supposed tellurate of mercury, was first described in ores of the American (Smith, 1883; Eckel, 1961) and Keystone (Genth, 1877) mines at Magnolia, but this mineral has also been discredited. A specimen labelled "magnolite, Keystone mine, Magnolia" and matching Genth's original descriptions was given to the writers by J. E. Byron of Boulder and this proved on X-ray analysis to be gypsum. A mineral corresponding to synthetic mercury tellurate (Christie, 1962) was sought in several samples of partially oxidized coloradoite, but without success.

Many new tellurite minerals have been discovered in recent years (Mandarino and others, 1961, 1963a, b) but the authors have been unable to find these in the present collection. Unfortunately, very few good specimens of the oxidized ores in which these minerals would most likely occur were available for study.

PHOSPHATE

Horn quartz of the tungsten veins contains abundant disseminations of a phosphate mineral variously identified as hamlinite (Hess and Schaller, 1914), goyazite (Schaller, 1917; Lovering, 1941), and woodhouseite

(Bonorino, 1959, p. 65). The mineral occurs as tiny, pale yellow, pseudocubic crystals of low birefringence scattered through quartz in amounts of from 10 to 25 percent (Lovering and Tweto, 1953, p. 48). This mineral is responsible for the green color of early quartz in the tungsten deposits.

This mineral also occurs in telluride ore of the Pride vein in the Poorman mine. Here it appears as abundant, weakly anisotropic, cuboid crystals 10 to 20 microns in size disseminated through an unusual variety of greenish-yellow horn quartz that coats the telluride-bearing quartz. Attempts to obtain pure samples of this mineral by heavy liquid separations were unsuccessful, but flame tests of the concentrates failed to indicate strontium. This would support identification as hamlinite or woodhouseite rather than goyazite.

Quantitatively, this phosphate is of no importance whatever in the telluride veins, but it may provide a significant link in the dating of the telluride and tungsten ores; though it appears late in telluride ore of the Pride vein, it precedes or accompanies early ferberite in the tungsten ores (Lovering and Tweto, 1953, p. 48).

TUNGSTATES

Ferberite

As the chief ore mineral of the tungsten veins, ferberite normally occurs in deposits younger than and separate from the gold tellurides. However, both types of ore occur in some veins of the Gold Hill and Magnolia districts, as previously mentioned, and both types have actually been exploited in several mines such as the Logan, Kekionga, and Red Signe. Very minor and local quantities of ferberite also formed as a product of the telluride stage of mineralization but these are commercially unimportant.

Lovering and Tweto (1953) and Lovering and Goddard (1950) have previously summarized the field and microscopic evidence bearing upon the age relations of the tungsten and telluride ores. They conclude that minor ferberite deposited during the telluride stage of mineralization preceded the tellurides, but that the economically significant stages of tungsten mineralization followed deposition of the gold telluride ores. Their published observations are summarized in Table 8.

Specimens from the Grandview, Herald, and Kekionga mines containing ferberite were examined in the present study, and our observations with regard to paragenesis of this mineral concur with those cited above. The age relations are very well illustrated by Kekionga ores in which crystals of early ferberite are corroded by younger petzite and native gold (Pl. 12a), while abundant later ferberite intergrown with quartz coats the older ferberite-telluride intergrowth.

In all specimens examined, the ferberite occurs as well-formed crystals up to .7 x 2.0 mm long. Narrow, chisel-shaped crystals are most common (Lovering and Tweto, 1953), but other habits appear and have been described by Hess and Schaller (1914).

TABLE 8. PREVIOUSLY REPORTED ASSOCIATIONS AND PARAGENESIS OF TELLURIDES AND FERBERITE (FROM LOVERING AND GODDARD, 1950, AND LOVERING AND TWETO, 1953)

Mine	Paragenesis
Grand Republic	Ferberite predates pyritic quartz similar to that carrying tellurides
Grandview	Vugs in ferberite filled by various tellurides
Graphic	Ferberite perched on sylvanite crystals
Herald	Ferberite a product of gold telluride mineralization and definitely earlier than the tellurides
Kekionga	Ferberite principally later than sylvanite although some is contemporaneous
Logan	Tungsten ores thought to be later than the tellurides
Red Signe	Ferberite veins cut gold ore and ferberite cements fragments of gold telluride ore

In addition to previously mentioned localities, the writers have found minor ferberite in ore of the Poorman mine in which it occurs as needles and prisms in quartz that also contains younger sphalerite, chalcopyrite, and galena in vugs. The Poorman mine is approximately three miles east of, and well removed from, the main tungsten district.

Scheelite

Scheelite occurs in many veins of the tungsten district (Tweto, 1947; Lovering and Tweto, 1953), but is rarely associated with the telluride deposits. Argall (1943) reported "small crystals and spots of high-grade scheelite" in the Croesus mine. According to Tweto (personal commun., 1966), scheelite occurs in the lower tunnel of the Red Signe mine which produced both gold telluride and tungsten ores (Lovering and Tweto, 1953, p. 193). All samples in the present collection of telluride ores were examined with ultraviolet light, but no scheelite was detected.

SILICATES

Adularia

Adularia occurs both as a wall rock alteration product and as a minor mineral disseminated through early vein quartz. The writers have identified this mineral in vein quartz from the Colorado, Melvina, Poorman, and Shirley mines and in all likelihood it could be found in small amounts in most of the veins.

Under the microscope, adularia appears as euhedral to subhedral rhombic crystals up to 30 microns in size that are either entirely surrounded by microcrystalline quartz or attached to altered wall rock along its contacts with vein quartz. Adularia is consistently an early gangue mineral that predates the metallic minerals with the exception of some pyrite-marcasite. Telluride seams commonly crosscut quartz containing adularia.

Chalcedony

Very little true chalcedony occurs in the telluride veins, and the term chalcedonic horn quartz seen in some of the older literature is inaccurate. Typical "horn quartz" is very fine- to medium-grained, gray to black quartz. Sparse chalcedony does appear as rare cavity fillings in horn quartz and, under the microscope, displays a distinctive radial aggregate structure.

Opal

Opal, a common late mineral in the tungsten veins (Lovering and Tweto, 1953, p. 44), is rare in the telluride deposits. Most of the milky translucent silica suspected of being opal proved on thin section study to be microcrystalline quartz.

Minor amounts of opal occur as very spotty vug fillings in horn quartz of the Gladiator, John Jay, Shirley, and White Crow mines. Opal in the John Jay ore is pale blue-white and displays rhythmic banding of the Liesegang type normally formed in gels. In specimens from the Grandview mine, films of opal coat both telluride-bearing and younger ferberite-bearing quartz. Opal is also intergrown with traces of chalcedony in ores of the Buena mine and both minerals cement a microbreccia of finely crushed granite. The ore minerals formed by replacement of this breccia prior to its cementation by the late opal.

Quartz

The ore minerals are invariably accompanied by quartz which, along with altered wall rock, comprises the bulk of vein material. Most of the veins are actually an interlacing network of horn quartz seams which are individually 1/16 inch to 3 inches wide (Pl. 1). In many of the veins, these seams vary greatly in thickness, number, and continuity and hence the ore was difficult to follow.

Quartz was evidently stable throughout the period of ore deposition and is represented by a complex sequence of pre- and post-telluride varieties, ending with sparse chalcedony and opal in the final stages of mineralization. Slight variations in texture, coupled with differences in the nature and abundance of included gangue and metallic minerals, produce great variability in the appearance of quartz from place to place in the veins. Several distinct generations are normally present in any ore sample and can often be recognized in other samples from the same deposit, but no meaningful correlation can be made from mine to mine.

Pyrite and marcasite are the most abundant contaminants of vein quartz and are primarily responsible for the dark gray color of typical "horn." Under the microscope myriad grains of pyrite are abundantly disseminated through most of the vein quartz.

Vein quartz ranges in size from coarse single euhedra 1 cm long down to grains that can hardly be resolved under the microscope, but most of the quartz is in the range .005 to 1.0 mm. In general, there is a trend toward finer grain sizes in later generations, possibly reflecting a progressive drop in growth temperatures.

The majority of fine quartz grains in typical horn are tightly intergrown and at best display subhedral forms. However, euhedral forms are common in early quartz that grew prior to crowding of available openings and also in late quartz along the borders of vugs.

Quartz does not appear to have been extensively replaced by the younger ore minerals and commonly displays euhedral outlines against both the sulfides and tellurides (Pls. 3g, 4d and 4g). Minor rounding, pitting, and local replacement of quartz are seen but did not play an important role in emplacement of the ore minerals (Pl. 12b). In contemporaneous intergrowths with quartz, early minerals like pyrite, marcasite, and ferberite commonly developed perfect forms (Pls. 4a, 4c and 12a), whereas the tellurium minerals are typically subhedral or anhedral (Pls. 2g and 13a). Late quartz which followed the ore minerals generally conforms to the outlines of even the most fragile metallic crystals (Pls. 5c and 7g).

As implied in the preceding paragraph, quartz completely overlapped deposition of the tellurides and native metals, but by far the bulk of the quartz formed before introduction of the tellurides. Hence, the ore minerals are most abundant as vug fillings and in seams transecting older quartz.

Telluride-bearing quartz varies greatly in color and texture, and there are no few types that can be singled out as significant hosts. However, the coarsest, glassy varieties and several late generations of extremely dense, fine-grained quartz with either a white, tan, or light green color are decidedly unfavorable. The compact, barren varieties are unusually hard and difficult to break in contrast to the more porous quartz of intermediate grain sizes that normally contains the tellurides, and this was early recognized by the miners.

The contacts of vein quartz with altered wall rocks are sharp and fairly straight and the majority of seams are best interpreted as simple fracture fillings. In some ores of the John Jay and King Wilhelm mines, however, telluride-bearing horn occurs in very irregular pockets which, at first glance, appear to have been produced by replacement of altered granite. A different origin is indicated by the presence of narrow layers of comparatively coarse, clear subhedral quartz along the borders of the horn quartz; these quartz crystals project into the pockets and appear to be early linings of irregular openings in the altered wall rock. This unusual type of ore may have resulted from hydrothermal leaching of the altered wall rock along irregular fractures prior to introduction of the horn quartz.

Roscoelite

Since its early recognition in ores of the Magnolia district (Genth, 1877; Endlich, 1878), roscoelite has been reported as a common constituent of the telluride ores (Lindgren, 1907; Goddard, 1940; Lovering and Goddard,

1950). Based upon its very close association with the gold tellurides and native gold, green roscoelite-bearing quartz was sought by the miners as a guide to ore. In 1910, some ore was shipped from the Kekionga mine of the Magnolia district for its vanadium content, probably in the form of roscoelite (Lovering and Goddard, 1950, p. 234). Some of this ore was reported to contain as much as 6.28 percent vanadium oxide and a number of samples tested contained an average 4.3 percent. A moderate tonnage of 2 percent ore was blocked out.

The older reports of roscoelite in Boulder County have never been systematically tested or even confirmed by X-ray methods (see Heinrich and others, 1953). For this reason, the writers selected a number of roscoelite (?) samples from different mines for detailed study; these were all very dark, emerald-green micas intergrown with quartz and the tellurides. Eleven samples were tested for vanadium and all gave positive results with an H_2O_2 test (Short, 1940) and a bead test described by Axelrod (1946). Five of these samples were X-rayed and gave patterns identical to one another and to published standards for roscoelite (Heinrich and others, 1953; Heinrich and Levinson, 1955). These results dispelled our concern that the reported roscoelites (?) might merely have been vanadian muscovites as suggested by the work of Heinrich and others (1953).

Roscoelite of the type tested here was previously reported in ores of the Gladiator, King Wilhelm, and Keystone mines and, in the present study, was found in ores of the Buena, Bumble Bee, Black Rose, Croesus, Emancipation, Gladiator, Keystone, King Wilhelm, Logan, Nancy, New Rival, Poorman, Success, Shirley, and Osceola-Interocean mines.

Roscoelite is an early vein mineral that formed along with pyritic quartz before deposition of the tellurides. In many specimens the roscoelite forms pure coatings up to 1 mm thick on altered wall rock or on early linings of coarse quartz along the borders of the telluride veinlets (Pl. 8c). Breccia fragments of wall rock or early vein matter were in places coated by roscoelite and later cemented by horn quartz. In contemporaneous intergrowths with quartz (Pl. 13b), roscoelite forms irregular clots, scalloped crusts, and radiating aggregates which are in places veined by the quartz. Where the tellurides appear in such intergrowths, they are markedly selective in replacing the roscoelite (Pl. 13b).

Precipitation of roscoelite ceased at about the time that sphalerite was beginning to form. In most specimens, sphalerite occupies cavities or seams in roscoelite-bearing quartz but in some ore of the King Wilhelm and Nancy mines roscoelite coats grains of early, anhedral sphalerite. The younger sulfides, tellurides, and native metals are consistently later than roscoelite.

The most abundant occurrence of roscoelite is in high-grade ore of the Walker-Clarke stope in the Buena mine. Here, rich and colorful intergrowths of green roscoelite, purple fluorite, quartz, and the tellurium minerals filled numerous fractures in a wide zone of shattered and altered granite. The roscoelite forms rosettes up to 2 mm in diameter which are mounted on early vein quartz and overgrown by later tellurides, fine-grained quartz, and clay. The small rosettes were apparently contemporaneous with cubes of purple fluorite. A few cubes of late, pale blue fluorite occur in late openings in this unusual ore.

OXIDES

Goethite

Microcrystalline goethite is the principal constituent of the "limonite" which occurs as abundant stains and coatings in the near-surface ores, and is derived chiefly from oxidation of fine-grained pyrite and marcasite in the primary ores. Much of the gold extracted from the outcrops and shallow underground workings was in the form of extremely fine spongy gold mixed with limonite, forming an intergrowth that the miners called "rusty gold." In many samples, the gold is too fine to be seen, but its presence is revealed by scratching the limonite which produces a golden metallic streak.

Goethite is commonly associated with jarosite and tellurium oxide in partially oxidized ores, but these minerals tend to be leached in thoroughly oxidized ore and hence are comparatively uncommon in material right at the surface. Wherever associated, goethite and jarosite show the curious zoned relationships previously discussed under "jarosite."

Poorly developed limonite boxworks after sphalerite and chalcopyrite are seen in a few samples, but the writers have never found boxworks that could be clearly related to original tellurides. This is due in large part to the fine-grained, intergrown and disseminated character of these ore minerals and also to their oxidation in the presence of large quantities of pyrite that favor transportation of the iron oxides.

Goethite is abundant only as a supergene mineral, but minor quantities in both altered wall rocks and vein quartz are hypogene. For example, in one sample from the Poorman mine, goethite and lesser hematite form rims and veinlets that surround and penetrate large feldspar crystals in altered granite adjacent to a quartz-telluride veinlet. These iron oxides could be related either to the breccia reef or the telluride stages of mineralization. Numerous subhedral crystals of pyrite are scattered through both the granite and vein quartz and show no signs whatever of oxidation. This sample comes from the 525-foot level of the mine several hundred feet below the deepest recognizable effects of weathering.

In some ore of the Smuggler mine, dense layers of limonitic horn alternate with unoxidized quartz-pyrite and carry unoxidized sylvanite-hessite in late vugs. The limonite pigment appears in rhythmic bands of the Liesegang type commonly developed in gels. Similar occurrences of hypogene iron hydroxide were noted by Lovering and Tweto (1953, p. 43) in the tungsten ores and, as discussed in later pages of this report, are significant in assigning temperatures of deposition to the ores in which they occur.

Hematite

Some hematite, both of the earthy and specular varieties, occurs in the telluride veins. Traces of hypogene hematite occurring in altered pyritic wall rock of the Poorman mine were described in preceding paragraphs. Also, in places along the Sentinel vein of the Golden Age mine, veinlets of fine-grained hematitic quartz up to 2 mm wide cut across pyritized, sericitized granite and these barren veinlets are truncated by seams of comparatively clear telluride-bearing quartz. Pyrite within the telluride seams shows no signs of oxidation.

Minor amounts of hypogene specular hematite formed in the telluride stage of mineralization in ores of the Eclipse and Smuggler mines. In both localities, minute scales of specularite appear in telluride-bearing quartz where they are corroded and replaced by early pyrite.

Stains of earthy supergene hematite occur locally in oxidized ore, but in amounts greatly subordinate to goethite, the predominant ferric oxide.

Paratellurite

Paratellurite, tetragonal TeO_2, has been identified only in oxidized ore of the Last Chance mine, but this single occurrence warrants some note as this is apparently the second known locality for this comparatively new mineral. Paratellurite was first reported by Switzer and Swanson (1960) who listed the type locality as either the Santa Rosa or La Moctezuma mine near Cananea, Sonora, Mexico. Subsequently, Mandarino found paratellurite in association with a number of new tellurites in La Moctezuma ores (Mandarino and others, 1961, 1963a and b).

As listed in Table 9, all lines in the powder pattern of the Last Chance material correspond in position and intensity to the published standards for paratellurite (Switzer and Swanson, 1960).

Paratellurite of the Last Chance ore occurs as bladed to anhedral grains up to 0.2 mm long replacing hypogene coloradoite and is in places intergrown with rickardite. In polished section, the paratellurite shows intense white internal reflections, is strongly bireflectant, and, for a gangue mineral,

TABLE 9. X-Ray Diffraction Data of Paratellurite*
Last Chance Mine, Jamestown District

d (Å)	I (visual)	d (Å)	I (visual)
4.03	1	1.53	3
3.41	90	1.49	10
2.99	100	1.41	1
2.41	10	1.27	7
2.08	3	1.23	5
2.02	3	1.19	5
1.88	70	1.12	1
1.71	3	1.09	3
1.69	15		

*Two hours exposure, CuKα radiation, Ni-filter, 57.3 mm camera.

is highly reflective (ca. 10 to 15 percent). Unfortunately, we do not have enough samples of the oxidized ores to determine the distribution and abundance of this mineral.

Tellurite

Tellurite, orthorhombic TeO_2, is common in partially oxidized ores, where it occurs as yellow stains and coatings closely associated with relict tellurides and with native tellurium. Crudely formed pseudomorphs of tellurite after sylvanite were found in dump samples of the Poorman mine. In many polished sections, fine-grained intergrowths of tellurite and gold are seen encroaching upon the primary gold tellurides (Pls. 2a and 9c). These gold-tellurite intergrowths are texturally identical to rusty gold also found replacing the tellurides (Pl. 12d) and in some cases may represent an intermediate stage in the development of rusty gold. Tellurite appears to be a temporary product easily leached from the oxidizing ores and, compared to limonite, is rare in the leached outcrops.

Other Oxides

Barren vein matter left in pits and trenches along the vein outcrops show local stains of manganese oxide, but these are uncommon. Films of manganese oxides also coat wire gold in oxidized ores of the Cold Spring-Red Cloud mine but have not been seen in association with other occurrences of supergene gold. The general rarity of manganese oxides relates to a paucity of primary manganese sources; rhodochrosite, the only known hypogene manganese mineral, is quite rare.

Lindgren (1907) reported blue stains on telluride ore samples on dumps at Eldora which he thought to be ilsemannite derived from abundant fine-grained molybdenite in the primary ores of that district. Lindgren's observations were evidently confined to the Eldora and Nederland districts, and did not extend to the main telluride belt. The present authors have seen no ilsemannite in the Jamestown, Gold Hill, or Magnolia districts, though traces of molybdenite do occur in the primary ores and, as previously noted, a small and unusual concentration of molybdenite was worked in the Mountain Lion mine at Magnolia.

UNIDENTIFIED MINERALS

Mineral A

An unidentified metallic mineral, possibly a new telluride of silver, is common in samples from the Gray Eagle mine situated midway between the Gold Hill and Jamestown districts. The properties of this mineral do not closely match those of any known species and are as follows:

APPEARANCE IN REFLECTED LIGHT:
Isotropic; Color, gray with tan tint tarnishing differentially blue-gray.

POLISHING CHARACTERISTICS:
Takes an excellent polish with few scratches; Talmadge Hardness A; H<chalcopyrite, gold, <<pyrite.

POWDER:
Black, opaque.

CLEAVAGE:
None observed; mineral somewhat brittle and breaks with an irregular to sub-conchoidal fracture.

INTERNAL REFLECTIONS:
None observed.

TWINNING:
None observed.

INDENTATION HARDNESS (Vickers):

Test Load	No. Tests	Mean H_v	Range H_v	
30 gm	26	56.4	39.6– 86.2	Excellent mark with
100 gm	12	100.3	90.8–100.6	little distortion

REFLECTIVITY:
White light, 3600°K.
Mean of 20 readings=22.8 percent
Range 20.5 to 24.0 (variable composition and polish)
Bireflectance Nil

POLARIZATION FIGURE:
In air, DR_r=V>r, very weak, DE Nil
In oil, DR_r Nil, DE Nil

ETCH REACTIONS:
$FeCl_3$, immediate reaction, surface turns iridescent to black
$HgCl_2$, rapid reaction, surface turns iridescent to black
HNO_3, HCl, KOH, KCN negative for 60 seconds

MICROCHEMICAL TESTS:
Positive; Ag and Te
Uncertain; Fe and S (possible pyrite contamination)
Negative; As, Bi, Cu, Co, Hg, Mn, Ni, Pb, Se, Sb, Zn

X-RAY FLUORESCENCE:
Confirms silver and tellurium as major components with variable traces of copper.
Composition close to that of hessite, Ag_2Te.

STRONGEST X-RAY LINES:
(Cu/Ni), 114.6 mm diameter camera
2.21 (very strong)
2.10 (strong)
2.15 (moderate)
2.60 (moderate)

OCCURRENCE AND ASSOCIATIONS:
Dense, anhedral veinlets and vug-fillings in fine-grained vein quartz. The mineral is molded against older pyrite and is veined by both chalcopyrite and native gold.

Additional study, now underway, is needed to firmly establish the identity of this unknown. The mineral is probably a new species related in some way either to the isotropic silver-copper telluride reported in the Vindicator mine at Cripple Creek (Loughlin and Koschmann, 1935, p. 296) or to a gamma phase ($Ag_{1.9}Te$) produced synthetically by Kiukkula and Wagner (1957) and Cabri (1965b).

Mineral B

The authors have been unable to identify a sulfosalt found in several samples from the Croesus, King Wilhelm, and Little Johnny mines. The mineral is moderately bireflectant (yellow-white to violet-white), strongly anisotropic with blue-gray to yellow-white polarization colors, and displays occasional red internal reflections. The polish is variable but generally poor and the mineral has a deceptively dull, sooty appearance at low magnifications. At magnifications over 500 X, individual grains and fibers display high reflectivity. The mineral occurs as irregular to round fibrous clots up to .12 mm in diameter scattered through telluride-bearing quartz and, in the King Wilhelm ores, also appears as fibrous, microscopic linings of vugs occupied by younger petzite, coloradoite, and nagyagite (Pls. 6b and 9a). This unknown mineral is light sensitive and develops a variable iridescence if long exposed to the beam of a high power objective. Difficulties in identifying this material arise from its fine grain, poor polish, and questionable homogeneity. Qualitative electron probe analyses indicate the presence of lead, antimony, silver, and sulfur. Five separate attempts at X-ray analysis failed to produce acceptable powder patterns, but strong lines appearing in all of these patterns are 3.50, 3.35, 2.88, and 1.98 Å, which are most closely comparable with those of fizelyite (Berry and Thompson, 1962, p. 156).

Weissite(?)

Traces of a light-gray, moderately anisotropic supergene mineral are intergrown with rickardite, tellurite, and free gold in partially oxidized telluride ore from the Bondholder, Grouse, Horsefal, Poorman, and Potato Patch mines. Its identification as weissite(?) is very tentative and based only upon its associations with rickardite and tellurite and its optical properties as given by Thompson (1949, pp. 357-358). The mineral does not occur in quantities sufficient for X-ray analysis or even for reliable microchemical testing. The weissite(?) is most common as an alteration product of the gold tellurides, but in ore of the Potato Patch mine it forms minute veinlets replacing native tellurium along grain boundaries and in advance of wholesale replacement by tellurite.

Summary of Hypogene Associations

GENERAL REMARKS

In the preceding descriptions of the vein minerals only the outstanding hypogene associations and antipathies were mentioned without any systematic effort to record all associations observed. Such information will now be presented in condensed tabular form with separate attention given to mineral pairs that occur in contact, to the frequency of occurrence of the different associations, and to the distinction of pairs that are randomly and nonrandomly associated. The present summary is restricted to various possible two-phase associations, but in later pages multiphase associations of minerals in the system Au-Ag-Te are treated in detail. Associations known to involve supergene minerals are not included in the compilations presented here.

PAIRS IN CONTACT

Of prime importance is the record of those minerals which do and those which do not occur in mutual contact. This record reflects restrictions of bulk composition on the Phase Rule operative during ore deposition. In Table 10 all mineral pairs found in direct contact in at least one polished section are indicated. The record includes all possible associations of the 16 principal ore minerals as well as those of ten selected minor metallic minerals. With the exception of krennerite, all of the principal ore minerals have been studied in at least 40 polished sections and thus, for them, all existing associations have probably been discovered. In later pages, this record will be used in comparison of natural and synthetic phase relations in the system Au-Ag-Te, and it will prove helpful in ascertaining the degree of equilibrium attained during ore deposition.

FREQUENCIES OF ASSOCIATIONS

The following record (Table 10) of pairs in contact gives no indication of mineral pairs that are frequent and contribute substantially to bulk composition of the ores. Observed frequencies of occurrence for various combinations of the 16 principal ore minerals are also presented in Table 10. The numbers given are the percentages of 156 polished sections found to contain

111

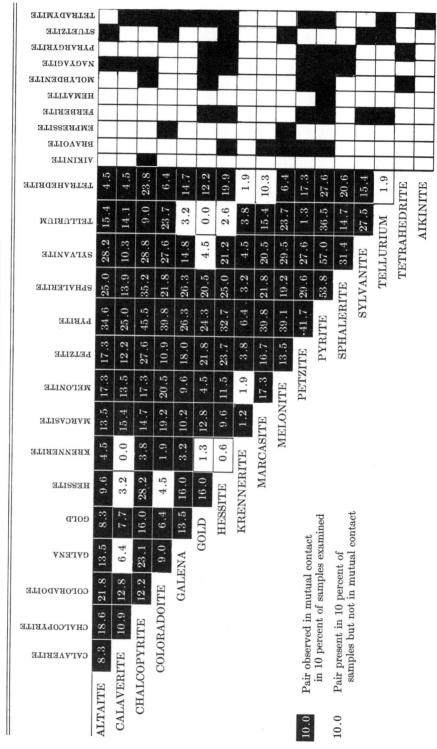

TABLE 10. SUMMARY OF HYPOGENE ASSOCIATIONS

the specified pair, but not necessarily in direct contact. It is immediately apparent from these data that a very large number of mineral pairs are moderately frequent in the Boulder ores and that these deposits cannot be adequately described in terms of a few predominant associations. With the exception of pairs involving pyrite, no single mineral combination was found in as many as one-third of the polished sections investigated.

PREFERENTIAL ASSOCIATIONS AND ANTIPATHIES

It is important to distinguish between randomly associated mineral pairs whose frequency of occurrence is governed merely by the abundances of the minerals involved, and those pairs whose frequency of occurrence requires a preferential association or disassociation. In some cases, especially where abundant minerals like sylvanite and gold never occur together, a biased relationship is obvious and requires no testing. In many cases, however, the distinction between random and nonrandom associations is more subtle and cannot be made simply by inspection of the data presented in Table 10. Consider, for example, the associations pyrite-gold and hessite-petzite which both occur in about one quarter of all samples analyzed. There is no apparent reason for suspecting the randomness of these associations. If random, however, then each pair should occur with a frequency about equal to the product of the frequencies of the individual minerals making up that pair. Since pyrite occurs in 96 percent and hypogene gold in 26 percent of all samples examined, these two should occur as a pair in about 25 percent (.96 x .26 x 100) of the samples. The observed frequency of this pair is 24 percent and very close to the expected value, so there is no basis for questioning the randomness of this particular association. For hessite-petzite, the expected frequency is only 16 percent, but the fact that this pair was actually observed in 24 percent of the polished sections suggests a positive, nonrandom association. Standard Chi-square tests can be applied to determine the probability of the discrepancy between observed and expected frequencies of each pair arising by chance alone. In the case of hessite-petzite discussed above, the determined discrepancy could arise only once in one hundred times (p = .01) by chance alone and is therefore judged significant.

Each of the 120 possible pairs of the 16 principal ore minerals were tested in this fashion, and the results are presented in Table 11. The numbers given indicate the probability levels for each pair, and those associations, both positive and negative, having a probability equal to or less than .20 are here considered significant.

While lending statistical support to suspected preferential associations, these tests revealed associative trends that went unnoticed during the microscopic studies. They show, for example, that pyrite, which associates frequently with all other metallic minerals in the Boulder ores, is not preferentially associated with any one in particular. Similarly, marcasite associates

TABLE 11.　SIGNIFICANT PREFERENTIAL ASSOCIATIONS AND ANTIPATHIES

	CALAVERITE	CHALCOPYRITE	COLORADOITE	GALENA	GOLD	HESSITE	KRENNERITE	MARCASITE	MELONITE	PETZITE	PYRITE	SPHALERITE	SYLVANITE	TELLURIUM	TETRAHEDRITE
ALTAITE			+ <.10		− <.01								+ <.10		− <.02
CALAVERITE						− <.01	− <.10	+ <.10					− <.20	+ <.10	− <.20
CHALCOPYRITE			− <.05	+ <.01		+ <.01				+ <.20		+ <.20		− <.01	+ <.01
COLORADOITE					− <.01	− <.01							+ <.01	+ <.01	− <.10
GALENA					+ <.01	+ <.05				+ <.10		+ <.01		− <.01	+ <.01
GOLD						+ <.01			− <.05	+ <.01		+ <.20	− <.01	+ <.01	
HESSITE							− <.20								
KRENNERITE								− <.20		+ <.01				+ <.01	
MARCASITE															
MELONITE										− <.20		− <.20	+ <.10	+ <.01	− <.10
PETZITE												+ <.01	− <.01	− <.01	+ <.01
PYRITE															
SPHALERITE														− <.10	
SYLVANITE														+ <.10	+ <.20
TELLURIUM															− <.01

frequently and randomly with most other vein minerals, but, for reasons unknown, tends to occur preferentially with calaverite and is significantly antipathetic to hessite and krennerite. The sulfides, tetrahedrite, chalcopyrite, galena, and sphalerite, are not only preferentially associated with each other but also show a significant affinity for native gold and its common associates, hessite and petzite. These minerals are correspondingly lean in ores that contain free tellurium. The strong coloradoite-gold association, important in the Kalgoorlie deposits, is missing in the Boulder ores. Actually, coloradoite, along with melonite and altaite, is somewhat antipathetic to free gold and is significantly concentrated in samples bearing native tellurium. It may well be that the mercury telluride tends to affiliate with the gold-rich phase present in the most tellurium-rich ores of a given district. Such a tendency would explain not only its preference for free gold at Kalgoorlie (where native tellurium is rare), but also the preferred association with sylvanite in tellurium-rich ores of Boulder County.

A substantial number of the significant associations and antipathies of Table 11 involve pairs of minerals in the system Au-Ag-Te. These relations, as well as the record of those pairs found in contact, are for the most part in accord with experimental and natural tie lines determined by Markham (1960), and will be fully discussed in later pages.

The Assay Record

The grade of the telluride ores varied greatly depending not only upon the kinds and proportions of ore minerals present, but also upon the degree of hand-sorting performed in the mines. In the early mining years prior to 1900 the shipments were generally small and high grade, some having values up to $10 per pound. These small lots of ore consisted of crude hand-sorted material containing the horn quartz veinlets and ranged in weight from about 50 pounds to 7 tons, and averaged less than 2 tons. An excellent record of this early production is preserved in assay records of the Boulder Sampler[5], a milling and sampling works in Boulder which handled many high grade shipments from the telluride belt in the periods 1878 to 1889 and 1890 to 1892. This assay record is summarized in Table 13 which lists the mean gold and silver contents and average gold:silver ratios of over 1500 shipments of telluride ore from 46 different mines scattered throughout Boulder County. This record is further condensed in Table 12 which gives comparable information for the three principal mining districts, Jamestown, Gold Hill, and Magnolia. The tenor of these early shipments ranged from about 3 to 286 ounces of gold and from one to several hundred ounces of silver to the ton. After 1900, much of the ore, though equally rich, was shipped in larger tonnages which generally assayed 0.5 to 15.0 ounces of gold and 0.5 to 25.0 ounces of silver to the ton. The gold:silver ratios in the telluride ores, as judged from the Boulder Sampler record, averaged 0.62 for the Jamestown district, 0.49 for the Gold Hill district, and 1.13 for the Magnolia district. As evident in Table 12, the chief difference among the three districts is the comparatively low silver content of the Magnolia ores and the resulting high gold:silver ratios of ores mined in that center.

It is tempting to postulate causes for recorded differences of gold and silver in the three districts, but, unfortunately, the assay record gives no indication of the degree of oxidation or even the depths of occurrence of the shipments that were assayed. For this reason, it is difficult to weigh the influence of supergene processes as distinct from original hypogene

---

[5]Four of the original ledgers of this firm are in the private file of the Editor of the Boulder Daily Camera in Boulder.

TABLE 12.   GOLD AND SILVER ASSAY DATA OF HIGH-GRADE TELLURIDE ORES
FROM THE
JAMESTOWN, GOLD HILL, AND MAGNOLIA DISTRICTS

DISTRICT	JAMESTOWN	GOLD HILL	MAGNOLIA
Number of Shipments	242	841	465
Average Shipment Weight (Pounds)	2436	3833	1178
*Average Gold Content (Troy oz./ton)	12.98 ± 1.79	9.38 ± 2.04	10.20 ± 0.84
*Average Silver Content (Troy oz./ton)	21.24 ± 5.37	19.24 ± 1.84	9.00 ± 1.18
Average Gold: Silver Weight Ratio	0.62	0.49	1.13
†Correlation Coefficient, R, Gold versus Silver	$+ .720 \begin{cases} .775 \\ .653 \end{cases}$	$+ .449 \begin{cases} .501 \\ .393 \end{cases}$	$+ .145 \begin{cases} .233 \\ .055 \end{cases}$
Average Depth of Mines (Feet)	600	563	180

*Range indicated is Standard Error of the mean.
†Range indicated is asymmetric 95% Confidence Interval on R.

gold:silver variations. As shown in Figure 12, mine depths in the Magnolia district were quite shallow and averaged only 180 feet in comparison with average depths of 560 and 600 feet for the Gold Hill and Jamestown districts, respectively. Therefore, a much larger proportion of the Magnolia production came from the near-surface zone of partial oxidation which typically persists to depths of about 150 feet in all three districts. This fact, along with the known trend for selective removal of silver during oxidation, suggests that supergene processes probably played an important role in causing the low silver content of the Magnolia ores. On the other hand, if this were due only to supergene processes, it would require that about 50 percent of all gold mined in the Magnolia district be in the form of supergene native gold and, although the record is inadequate in this regard, it is clear that the proportions of supergene gold in this district were not nearly so large. The ores of the Magnolia district, as in the other districts, were predominantly hypogene with perhaps 10 to 15 percent as a more realistic estimate of the fraction of total gold that was mined in the form of rusty gold. In this connection, it is of interest that gold and silver are positively and significantly correlated in assayed ores from the Jamestown and Gold Hill districts, whereas no such correlation could be established for the Magnolia ores (see correlation coefficients in Table 12). This may reflect the greater proportionate influence of supergene leaching in this district, but the indications cannot be very rigorously interpreted.

In turning to gold and silver variations within the Jamestown and Gold Hill districts, it is safer to interpret these as hypogene effects because supergene ore comprised such small amounts in the total record of production.

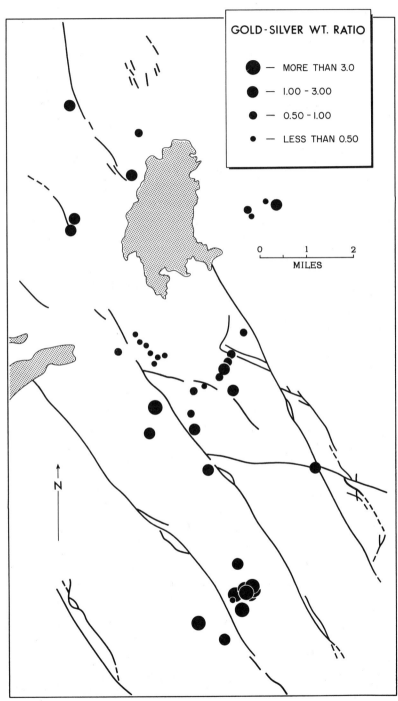

Figure 11. Map showing distribution of gold:silver ratios in telluride mines of Boulder County.

TABLE 13. GOLD AND SILVER ASSAY DATA OF HIGH-GRADE TELLURIDE ORES OF INDIVIDUAL MINES IN BOULDER COUNTY

PROPERTY	Number of Shipments Considered	Average Shipment (lbs)	Mean Gold Content (Troy oz/ton) *	Mean Silver Content (Troy oz/ton) *	Au:Ag Weight Ratio *
JAMESTOWN DISTRICT					
Atlantic	4	575	3.20 ± 0.61	1.78 ± 0.57	2.13 ± 0.45
Buena	36	2442	17.20 ± 6.00	6.46 ± 1.53	2.49 ± 0.24
Ellen	11	378	23.17 ± 10.7	46.14 ± 21.8	0.39 ± 0.07
Gladiator	6	144	82.03 ± 36.3	299.50 ± 163.9	0.66 ± 0.25
Hercules	7	169	2.90 ± 1.46	8.71 ± 5.00	0.62 ± 0.23
John Jay	60	3051	8.30 ± 0.94	2.78 ± 0.28	3.55 ± 0.26
King Wilhelm	10	179	22.88 ± 11.73	18.37 ± 8.95	1.38 ± 0.51
Last Chance	14	1829	11.03 ± 2.15	8.65 ± 3.50	1.88 ± 0.59
Monitor	23	1351	3.32 ± 1.62	36.81 ± 12.6	0.85 ± 0.49
Rip Van Dam	13	603	18.75 ± 11.5	58.81 ± 38.0	0.35 ± 0.04
Smuggler	52	4625	7.77 ± 1.02	7.40 ± 0.81	0.99 ± 0.05
GOLD HILL DISTRICT					
Alpine Horn	10	430	6.58 ± 1.51	3.92 ± 0.73	1.62 ± 0.17
American	44	2317	16.02 ± 6.50	25.90 ± 6.22	0.88 ± 0.14
Cash	56	4312	3.16 ± 0.21	24.24 ± 3.15	0.20 ± 0.02
Cold Spring-Red Cloud	47	6466	7.15 ± 1.27	15.86 ± 2.19	0.56 ± 0.09
Colorado	18	1142	6.16 ± 0.44	1.59 ± 0.18	4.47 ± 0.49
Emancipation	51	3986	5.74 ± 0.43	4.52 ± 0.36	1.36 ± 0.07
Franklin	21	2480	4.04 ± 0.44	11.01 ± 2.71	1.44 ± 0.45
Freiberg	21	1346	2.50 ± 0.37	8.94 ± 1.36	0.33 ± 0.04
Grandview	51	2733	4.83 ± 1.29	8.81 ± 1.02	0.55 ± 0.07

Horsefal	31	1146	3.08 ± 0.26	7.87 ± 0.65	0.42 ± 0.03
Ingram	50	4389	7.70 ± 1.33	20.73 ± 4.75	0.45 ± 0.02
Logan	95	1151	14.89 ± 6.52	26.13 ± 11.5	1.35 ± 0.26
Melvina	50	1646	5.63 ± 0.48	10.34 ± 1.38	0.92 ± 0.11
Nil Desperandum	5	884	9.76 ± 5.09	14.89 ± 3.96	0.90 ± 0.19
Osceola-Interocean	44	4536	4.40 ± 0.40	10.80 ± 1.09	0.52 ± 0.07
Poorman's Relief	18	3563	91.32 ± 84.5	76.00 ± 41.7	1.03 ± 0.28
Prussian	48	12931	9.82 ± 3.27	25.40 ± 8.63	0.48 ± 0.04
Richmond	18	2231	4.91 ± 2.11	10.17 ± 2.42	0.54 ± 0.09
Slide	50	9496	9.42 ± 0.67	31.11 ± 2.27	0.31 ± 0.01
Sterling	37	1187	4.04 ± 0.32	17.07 ± 1.96	0.39 ± 0.04
Temborine	3	283	3.80 ± 1.04	2.80 ± 1.11	1.73 ± 0.65
MAGNOLIA DISTRICT					
Acme	14	272	10.49 ± 5.75	7.23 ± 3.40	1.20 ± 0.21
Eclipse	10	147	6.94 ± 4.33	3.26 ± 0.68	2.00 ± 0.50
Fortune	15	871	0.71 ± 0.13	104.47 ± 20.7	0.01 ± 0.003
Graphic	15	966	5.42 ± 1.67	1.76 ± 0.39	3.77 ± 0.62
Kekionga	48	3420	9.18 ± 2.31	1.83 ± 0.25	4.54 ± 0.69
Keystone	51	678	10.80 ± 1.38	2.95 ± 0.30	4.82 ± 0.70
Lady Franklin	54	675	11.91 ± 2.32	1.67 ± 0.16	5.78 ± 0.58
Little Maud	7	382	6.44 ± 2.23	1.33 ± 0.19	4.47 ± 0.50
Little Pittsburgh	54	2144	7.44 ± 2.10	23.30 ± 10.20	0.46 ± 0.05
Mountain Lion	59	611	11.54 ± 1.33	3.81 ± 1.39	3.54 ± 0.32
Ophir	43	373	10.31 ± 1.93	8.64 ± 3.18	5.11 ± 0.69
Pickwick	18	2013	15.15 ± 8.17	1.16 ± 0.22	8.01 ± 2.20
Sac and Fox	29	460	10.88 ± 1.46	3.47 ± 0.54	4.18 ± 0.51
Senator Hill	48	1238	13.31 ± 5.10	2.98 ± 0.69	3.98 ± 0.63

*Range indicated is standard error.

In general, gold:silver ratios varied irregularly from mine to mine within these two districts. As revealed in Figure 11, the Cash and Slide centers of the Gold Hill district stand out as areas of unusually low Au:Ag ratios more or less centrally located with respect to the entire telluride belt; otherwise there do not appear to be any notable trends of Au:Ag distributions.

Figure 12. Elevations of mine workings in representative telluride mines of the Jamestown, Gold Hill and Magnolia districts.

Distribution of Minerals

LOCAL DISTRIBUTION

The individual telluride veins were never systematically sampled and analyzed from the standpoint of possible mineralogic zoning, but the impression gained by those who have mapped the deposits underground (Goddard, 1940; Lovering and Goddard, 1950) is one of an irregular distribution of the ore minerals. The miners themselves, who were keenly aware of certain features such as the proportions of sulfides, tellurides, and free gold, found these changes to be spatially inconsistent and unpredictable. Ore mined over a vertical range of 1000 feet in some of the larger mines like the Slide and the Ingram was essentially unzoned and showed no definite trends of bottoming in sulfides or barren pyritic quartz.

In a brief commentary on the Boulder telluride deposits, Wahlstrom (1950, p. 949) states that "in many mines the telluride ores grade downward or laterally into lead-zinc-silver ores and show a crude zonal relationship." Unfortunately, he does not give specific mine examples and the present authors do not know of a single example where such a gradation has been documented. It is true that the quantities of sphalerite and galena formed in the telluride stage of mineralization varied among the mines and in individual deposits, but the lead and zinc were neither commercial nor regularly distributed with respect to the tellurides. The telluride ores contained atypically large amounts of base metal sulfides in a few mines such as the Cash and Colorado-Rex, where there has been some exploration, stockpiling, and minor gold-silver production in recent years. Even in these deposits, however, no zonal gradation between telluride ore and lead-zinc-silver "ore" has been demonstrated. Most of the commercially important lead-silver ores in the telluride belt are distinctly older than the telluride ores (Goddard, 1940; Lovering and Goddard, 1950) and do not, therefore, intergrade with them.

REGIONAL DISTRIBUTION

One objective of the present study has been to determine any systematic zoning of the tellurides on a regional scale that might be related to major

123

centers of mineralization. The available polished section record (Table 7) has been used as a guide to the distribution of the ore minerals but not necessarily to their true abundances. The premise has been that any pronounced regional zoning involving the presence or absence of certain minerals in distinct zones would reveal itself in the mineralogy of numerous widespread samples even though the average mineral abundance at any one point might be unknown.

Maps were prepared for each of the ore minerals showing their recorded occurrences and average sample abundances, and ten of these maps are reproduced in Figure 13. It is immediately evident from these plots that all of the minerals shown (with the possible exception of krennerite) are widely dispersed in different deposits, and that the telluride belt cannot be subdivided into broad and continuous zones of distinctive mineralogy. Krennerite occurs infrequently but in substantial quantities; it is less widespread than the other tellurides and may give a vague impression of clustering around the Early Tertiary stocks shown in Figure 13. However, these stocks are older than and not genetically related to the mineralization and, even though they have probably had a structural influence on the distribution of the tellurides as previously mentioned, they should not be regarded as sources or centers of mineralization.

Although there are no large-scale patterns of regional zoning of the telluride minerals, their distribution does not seem to be entirely chaotic. As previously discussed, the belt is comprised of about 13 separate productive centers, and among those that have been studied in greater detail there are rather distinctive differences of bulk composition and mineralogy. The differences are not extreme but rather involve variations in the proportions or combinations of minerals that occur at least in minor amount in most of the veins. (Hence these differences are not well expressed in Figure 13 which gives equal weight to abundances determined by varied numbers of samples taken in the different mines.) For examples, there are the John Jay center with its unusual quantities of native tellurium and complete lack of free gold, the Smuggler center containing abundant altaite and free gold in direct contact—two abundant minerals that are widespread but cannot be found in contact elsewhere in the belt, and the Buena center whose ores can instantly be recognized as unique colorful intergrowths of abundant roscoelite, fluorite, tellurides, and native tellurium. The Cash and Slide centers, as noted before, are distinguished by their unusually high silver (hessite + petzite) and base metal sulfide contents. The fact that it is only in these centers that sphalerite and chalcopyrite are found in exsolution-type intergrowths suggests that temperature as well as compositional factors may have had some influence on the character of mineralization in these different centers.

It is suggested that the mineralogical differences among these centers arose chiefly from compositional variations in the fluids that mineralized each center, that these fluids inherited their differences in the source area at depth, and that the degree of structural isolation of each center was a

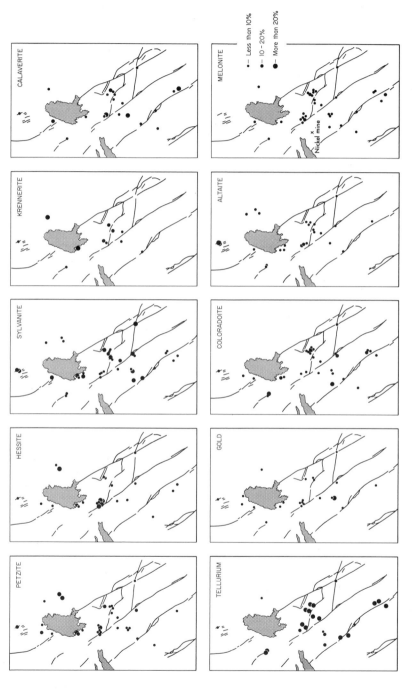

Figure 13. Maps showing the recorded occurrences of some significant tellurides and other ore minerals in the telluride mines of Boulder County.

factor in preserving the compositional identity of the ascending fluids and the uniqueness of the resulting mineral assemblages. Centers like the Buena, John Jay, and Smuggler are localized along single breccia reefs remote from other centers or reef intersections and were probably mineralized under structurally isolated conditions. Other centers like the Poorman (Fig. 7) situated at the intersection of two major reefs may have been mineralized by fluids ascending both structures and perhaps mixing along the way. The veins of the Poorman center vary greatly and, as a group, lack a distinctive composition and mineralogy of their own. In mapping the Poorman mine underground and in searching its dumps, the writers have found samples of telluride ore containing native tellurium, others with free gold, and still others with large amounts of sulfides as though this one center was a microcosm of the entire telluride belt. In later pages, the controls of mineralogy in these centers are further discussed, but it is important here to stress that there is this preferential occurrence of certain ore minerals in some of the centers and that this is the only degree of systematic mineral distribution in a belt that seems otherwise unzoned.

Paragenesis

THE GENERAL VEIN SEQUENCE

The paragenetic sequence of the telluride veins is complex and involves an unusually large number of hypogene minerals. Viewed from the standpoint of age relations of chief gangue minerals and groups of metallic minerals, the sequence is clear-cut and easily established, but detailed age relations among the individual metallic minerals, particularly the tellurides, are more difficult to determine. For this reason, attention is first given to the general sequence of gangue minerals and major metallic groups before turning to a more detailed treatment of age relations among the individual ore minerals.

The general vein sequence is presented in Figure 14, a paragenetic diagram based upon evidence from thin section, polished section, and hand specimen studies. This over-all sequence is pieced together from the fragmentary sequences of numerous ore samples, and is somewhat artificial in the sense that certain combinations of minerals listed (e.g., gold and tellurium) are incompatible and do not occur in direct contact.

Several features of the paragenesis are noteworthy. The vein mineralization commenced with quartz whose deposition continued spasmodically and diminishingly throughout the ore-forming process, giving way in final stages to sparse chalcedony and opal. Adularia, roscoelite, fluorite, and barite formed in that order and, as a group, preceded the tellurides and native metals. Several pulses or substages of sulfide deposition are indicated. The first two of these are predominantly pyrite-marcasite, the third is predominantly sphalerite, and the fourth is galena-chalcopyrite. Traces of chalcopyrite were contemporaneous with late aikinite and tetradymite, and the very last sulfide to be deposited was minor pyrite occurring as sparse crystals on late vein carbonates in some of the mines. The early ferberite shown in Figure 14 is that formed as an integral part of the telluride stage of vein formation and not during the main tungsten stage that followed in some of the mines. The timing of carbonate mineral deposition is interesting; some ankerite-dolomite preceded the tellurides, but the veins in general are characterized by a sudden appearance of ankerite-dolomite and calcite late in the sequence after the tellurides had formed.

127

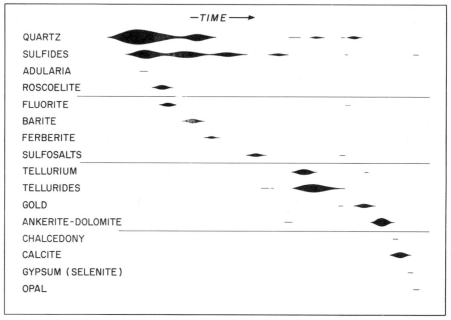

Figure 14. General paragenesis of the Boulder County telluride veins.

The writers have included chalcedony, opal, and selenite in the hypogene sequence, but cannot rule out the possibility that these late minerals are entirely supergene. Most gypsum found is supergene, but coarse selenite and also precipitates of opal and chalcedony found deep beneath the zone of weathering may have been deposited by late ore fluids.

The outstanding impression of the general sequence is that of a continuing process of mineralization uninterrupted by major structural events or even by significant lapses in deposition. There is no evidence of major reversals or repetitions of the sequence as shown in Figure 14. There is also sufficient overlap in deposition of the sulfides, tellurides, and native metals to indicate that all were formed as parts of the same general "wave" of mineralization.

Structural movements during mineralization were confined largely to fracturing that persisted after initial opening of the veins and permitted emplacement of numerous crosscutting seams of quartz and pyritic quartz. Brecciation and crushing of vein material locally affected early quartz and fluorite but ceased before mineralization was long underway. Fracturing of subdued intensity continued into the earliest stages of telluride deposition, but the tellurides are for the most part undeformed. Shearing and brecciation of native tellurium and the tellurides of the type noted in ores of the Alpine Horn (Pl. 7f) and White Crow mines (Pl. 3c) are truly exceptional features. Actually, part of the difficulty experienced in working out the age relations of the tellurides stems from the fact that they are not extensively fractured and, as a result, straightforward veining relationships are uncommon. Minor fracturing followed deposition of the tellurides as evidenced by late carbonate veinlets that typically crosscut ore, but such deformation did not seriously distort the older ore minerals.

THE METALLIC SEQUENCE

Detailed age relations among the ore minerals are shown in Figure 15. As in Figure 14, this is a diagram that shows the age relations of many minerals irrespective of the fact that certain combinations of these are chemically or environmentally incompatible. At the outset, it should be stressed that this sequence is by no means a simple and continuous succession of precipitates formed in response to a changing ore fluid. It combines, without distinction, partial sequences that occur separately in the veins. It also shows the known effects of cooling which have modified the mineralogy and textures of the initial assemblages to varying degrees. Where more than one generation of a given mineral appears in this sequence, this is noted on the diagram and explained in the following text. Many of the complex cooling reactions involving metastable phases indicated in Figure 15 deserve more

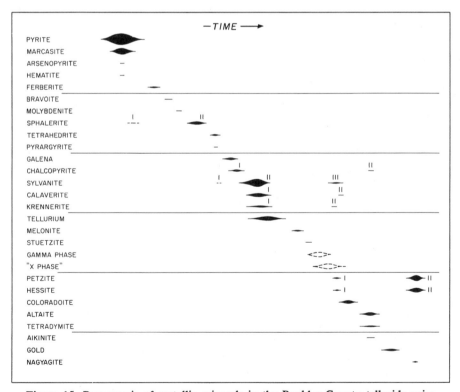

Figure 15. Paragenesis of metallic minerals in the Boulder County telluride veins.

than the passing comment they receive here; these are more fully discussed under the next heading, "Phase Relations of the System Au-Ag-Te."

Compared to later ore minerals, age relations among the early sulfides and sulfosalts are straightforward. The only complication introduced by cooling is possible local exsolution of chalcopyrite from sphalerite. It is difficult to know where to place such chalcopyrite in sequence with later ore minerals. Two generations of sphalerite are indicated. The first is represented by minor sphalerite found in the altered wall rocks bordering the telluride veinlets and by sparse sphalerite contemporaneous and intergrown with early pyritic quartz. However, most of the sphalerite is later and typically occurs in vugs in pyritic quartz where it is replaced by younger ore minerals. Chalcopyrite and galena appear contemporaneous in many samples, but where a difference in age is indicated, chalcopyrite is more commonly the younger of the two. Chalcopyrite I is that formed in the main stage of sulfide deposition while chalcopyrite II is the very rare variety formed in association with late aikinite and tetradymite.

Among the tellurides, sylvanite has had the most complex history and formed at several points in the sequence. Sylvanite I formed contemporaneously with tetrahedrite (Pl. 11a) and is veined by chalcopyrite I (Pl. 3e). Most of the sylvanite in the veins formed contemporaneously with, but separately from, early calaverite and krennerite and is represented in Figure 15 as sylvanite II. Sylvanite III is late material formed upon breakdown of a metastable phase, the gamma phase of Cabri (1965b), and appears in the ores as fine-grained micrographic inclusions in hessite (Pls. 7a and c). Also formed at about this time were minor amounts of sylvanite occurring in veinlets of petzite that crosscut sylvanite II, calaverite I, and krennerite I (Pls. 10g and 10h). Still another type of sylvanite was formed at this time (included under sylvanite III), and is represented by minute grains of sylvanite found in rims of coloradoite replacing native tellurium (Pl. 2f).

Most calaverite formed early in the sequence in advance of native tellurium but some (calaverite II) appears contemporaneous with coloradoite in intergrowths that replace native tellurium (Pl. 5c). Krennerite I is the main stage of krennerite deposition, but a trace of this mineral (krennerite II) formed late and under unusual conditions previously described for the ores of the Alpine Horn mine (Pl. 7f).

The timing of melonite with respect to native tellurium and the early gold tellurides has been discussed earlier. The placement of this mineral at an intermediate point in the paragenesis is a notable departure from Wahlstrom (1950) who considered melonite the earliest telluride to form.

Most of the native tellurium in the veins is contemporaneous with or slightly later than associated calaverite I, krennerite I, and sylvanite I and II, and is replaced by younger melonite, stuetzite, coloradoite, and altaite as well as by sparse sylvanite III and calaverite II. The relatively late tellurium II is the "sponge variety" found replacing several tellurides including late tetradymite (Pls. 3a and 3b).

There is convincing experimental evidence (Cabri, 1965b) that the hessite-petzite and hessite-sylvanite intergrowths seen so commonly in the Boulder County ores were derived from certain metastable phases that broke down upon cooling. The former existence of such phases complicates interpretation of the natural assemblages because it is difficult to determine exactly when they broke down with respect to the timing of other events late in the sequence. The stability relations and thermal history of these phases are discussed in detail under "Phase Relations of the System Au-Ag-Te" and, at this point, we shall merely indicate their probable positions in the paragenetic sequence. The metastable gamma phase breaks down at temperatures below at least 170°C, ultimately forming two- or three-phase intergrowths of hessite, petzite, stuetzite, and sylvanite. Not all the steps in this breakdown are known, but Figure 15 is drawn to show the probable formation of some sylvanite III, hessite I, and metastable "x" phase from the metastable gamma phase. The metastable "x" phase is stable down to at least 70°C, and below that temperature breaks down to hessite plus petzite. The "x" phase is believed to have formed in the Boulder County veins during and after deposition of the gamma phase, and persisted until the ores had cooled to below 70°C at which time it must have broken down to mutual intergrowths of hessite II and petzite II. Some hessite (I) and petzite (I) probably formed at temperatures above 70°C but not in stable contact with one another.

This problem will be dealt with later, but it should be stressed here that the true sequence of metallic minerals is not altogether evident from the textures seen under the microscope. For example, mutual intergrowths of petzite II and hessite II appear to be veined or replaced by many late minerals including altaite, tetradymite, aikinite, and native gold even though hessite and petzite cannot exist in direct contact at temperatures above $50 \pm 20°C$ (the lower stability limit of the "x" phase). The late altaite, aikinite, tetradymite, and gold very probably replaced the "x" phase which subsequently formed hessite II and petzite II late in the cooling process, perhaps long after hydrothermal fluids had ceased their flow through the veins.

Two of the numerous metallic minerals could not be placed with confidence in this sequence. Rare crystals of nagyagite occasionally seen in hessite or petzite probably exsolved from those minerals or possibly were exsolved from original "x" phase, and were preserved as the "x" phase unmixed to form hessite II and petzite II. In Figure 15, nagyagite is very tentatively shown as contemporaneous with hessite II and petzite II, but this is highly conjectural. Unfortunately, the mineral empressite was found only in contact with melonite and coloradoite and so its age relations are essentially unknown.

As a very generalized summary, the metallic sequence of the Boulder County veins was initiated by early sulfides, followed by a complex suite of tellurium minerals, and native gold was the latest ore mineral to form. Certain assemblages have been highly modified by cooling and the resulting textures and associations are deceptive if not interpreted in terms of known phase relations. Excluding these late effects, the initial sequence of tellurium minerals generally is a succession of phases of progressively lower tellurium content.

Phase Relations of the System Au-Ag-Te

EXPERIMENTAL PHASE RELATIONS

Many researchers have contributed to our present knowledge of phase relations in the system Au-Ag-Te, but the recent works of Markham (1960), Honea (1964), Cabri (1965b and c), and Kracek and others (1966) are most directly applicable to the interpretation of the Boulder County ores. Full reviews of previous experimental work on this system are given by both Markham (1960) and Cabri (1965b). Only some of the more critical

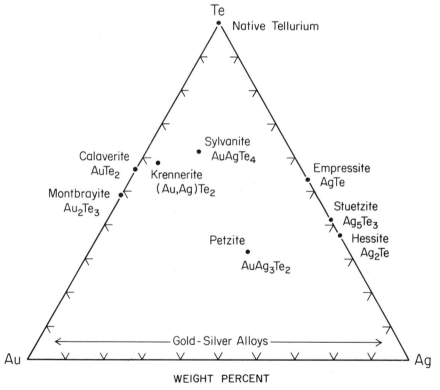

Figure 16. Minerals of the system Au-Ag-Te.

133

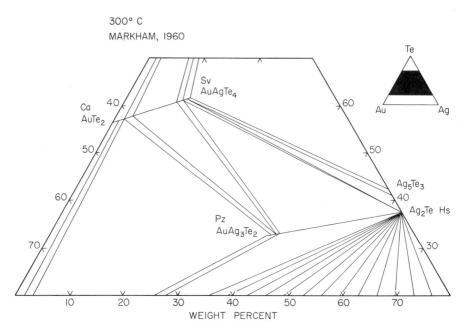

Figure 17. The 300°C isotherm of the synthetic system Au-Ag-Te according to Markham, 1960.

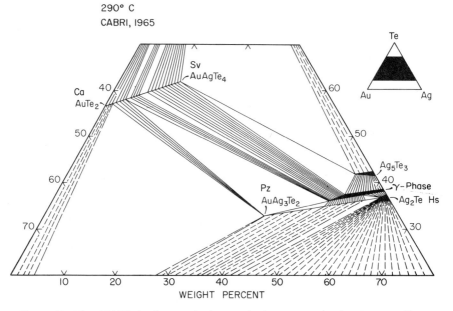

Figure 18. The 290°C isotherm of the synthetic system Au-Ag-Te according to Cabri, 1965b.

phase diagrams can be reproduced here, so the reader may wish to refer to the original works cited above.

There are nine valid mineral representatives of the system Au-Ag-Te (Fig. 16). With the exceptions of montbrayite (Au_2Te_3) and empressite (AgTe), all of these minerals have been synthesized in the laboratory and their stability relations are fairly well known. Montbrayite is extremely rare in nature and Tunell and Pauling (1952, p. 379) attribute its instability to geometrical difficulties in formulating an unstrained structure for the composition Au_2Te_3. Empressite (AgTe) is not altogether uncommon in telluride deposits, but experiments have failed to produce its synthetic counterpart (Honea, 1964; Cabri, 1965b). It should be stressed again that the writers are calling AgTe empressite and Ag_5Te_3 stuetzite following Honea's (1964) redefinition of these silver tellurides. There is apt to be some confusion here because the work of Markham (1960) preceded that of Honea and hence Ag_5Te_3 which Markham did synthesize and appropriately called empressite at that time, must now be called stuetzite.

Experimental tie lines of the system Au-Ag-Te as reported for 300°C by Markham (1960) and for 290°C by Cabri (1965b) are shown in Figures 17 and 18. Markham was unable to synthesize krennerite, and hence it does not appear in his experimental diagram. Obviously, these two investigators reported very different results for the same system at the same temperatures. With the exception of gold-silver alloys, Markham detected very limited solid solution among the phases of this system whereas Cabri reports extensive solid solution particularly among the silver-rich phases. Furthermore, Cabri recognized the existence of some phases unstable at room temperature, including the gamma phase ($Ag_{1.9}Te$) of Kiukkula and Wagner (1957) and a new phase, the "x" phase, stable along the join Ag_3AuTe_2-Ag_2Te and containing 2.5 to 14.5 weight percent gold.

These and other differences in the experimental results of Markham and Cabri probably stem from the techniques employed. Cabri used rapid quenching procedures and X-rayed synthetic products immediately after cooling. Unquenchable products were X-rayed with a high-temperature camera at the temperature of synthesis. He also used a Guinier-de Wolfe camera superior for the detection of minute amounts of any phase in a mixture. Markham does not mention special precautions of this kind and evidently some of his synthetic charges readjusted to low temperature between the time of synthesis and X-ray analysis. Phase relations portrayed by Cabri (Fig. 18) are very probably the true relations at 290°C, but those determined by Markham (Fig. 17) probably reflect tie lines that exist stably or metastably at much lower temperatures.

APPLICATION OF EXPERIMENTAL RELATIONS
TO THE BOULDER COUNTY DEPOSITS

General Remarks

In many respects, the Boulder County ores are ideally suited for analysis in terms of the synthetic system Au-Ag-Te. Unlike the ores of many other major telluride districts that are characterized by comparatively restricted bulk compositions, the Boulder ores vary greatly in the proportions of gold, silver, and tellurium (Fig. 20). As a result, useful experimental relationships pertaining to almost any part of the system can be applied to these ore deposits. All phases synthesized in the artificial system and stable at room temperature occur in one place or another in the veins. Also helpful is the fact that the deposits have not been extensively weathered, and so the hypogene relations are not seriously obscured by supergene effects.

Certain assumptions are inherent in any application of experimental equilibrium diagrams to hydrothermal assemblages formed in an open vein system and limitations peculiar to the system Au-Ag-Te are discussed by Cabri (1965b). It is assumed here that the Boulder ores formed in the presence of a vapor, and that the subsolidus relations applied are not significantly altered by the presence of H_2O or other volatiles. It is also assumed that the many components known to have accompanied gold, silver, and tellurium in the ore fluids did not substitute in solid phases of the system to the extent of changing the phase equilibria. These assumptions, particularly the last, may not be fully justified. No data are available on the trace element contents of the ore minerals and possible substitution of certain elements like selenium, sulfur, copper, and perhaps others in the gold-silver tellurides certainly cannot be ruled out. In view of these uncertainties, the only strong justification for correlating the natural and synthetic phase relations is the fact that the natural assemblages make a great deal of sense when analyzed in terms of the experimental work.

Cooling Versus Initial Deposition

The conflicting phase diagrams of Markham (Fig. 17) and Cabri (Fig. 18) are both useful in the interpretation of natural telluride assemblages. Cabri's research has established the equilibrium relationships to be expected in assemblages initially formed at elevated temperatures, whereas Markham has unintentionally provided a diagram showing the associations to be expected in ores readjusted to room temperature. There is, of course, no assurance that the tie lines drawn by Markham represent stable rather than metastable associations, but their close correspondence to natural associations suggests that they are at least the final products normally produced in the cooling of ores. Both diagrams have therefore been used in the present study in an effort to sort out features of the Boulder deposits that belong to stages of initial deposition, as opposed to later cooling effects that tend to mask the original assemblages. In the following pages, the record of associations of Au-Ag-Te minerals is first summarized and then this record is interpreted.

OBSERVED ASSOCIATIONS

The following combinations of minerals of the system Au-Ag-Te occur in mutual contact in the Boulder ores:

Calaverite-tellurium
Calaverite-sylvanite
Calaverite-petzite
Calaverite-AuAg alloy
Sylvanite-krennerite
Sylvanite-tellurium
Sylvanite-stuetzite
Sylvanite-hessite
Sylvanite-petzite
Stuetzite-tellurium
Stuetzite-hessite

Petzite-hessite
Petzite-AuAg alloy
Hessite-AuAg alloy
Krennerite-tellurium
Krennerite-petzite
Calaverite-petzite-sylvanite
Krennerite-petzite-sylvanite
Sylvanite-hessite-petzite
Sylvanite-hessite-stuetzite
Calaverite-petzite-AuAg alloy
Petzite-hessite-AuAg alloy
Sylvanite-stuetzite-tellurium.

With the exceptions of sylvanite associated with calaverite and stuetzite in contact with tellurium, all minerals have been X-rayed in the particular associations listed. No associations are listed for empressite because this mineral was found in only one sample and there contacts only melonite and coloradoite. All of the assemblages listed are evidently compatible with altaite, coloradoite, and most with melonite, but it is impractical to list here all conceivable multiphase associations of the 30 other metallic minerals with these phases of the Au-Ag-Te system.

Of particular interest are those Au-Ag-Te minerals which do not occur in contact, as these express either chemical incompatibilities or limitations of bulk composition. Associations in this category are as follows:

Calaverite-hessite
Calaverite-krennerite
Calaverite-stuetzite
Krennerite-hessite
Krennerite-stuetzite
Krennerite-AuAg alloy

Petzite-stuetzite
Stuetzite-AuAg alloy
Sylvanite-AuAg alloy
Tellurium-hessite
Tellurium-petzite
Tellurium-AuAg alloy.

With the exception of rare stuetzite, all of the minerals listed are sufficiently abundant that these associations would surely have been found if present in the veins. It is noteworthy that these apparently incompatible combinations account for a large number of the negatively associated pairs previously mentioned and listed in Table 11.

Figure 19 is a phase diagram based on the assemblages observed in the Boulder ores. For lack of chemical analyses of the gold, tie lines are drawn to the gold corner to reflect the estimated silver content (less than 10 weight percent) and follow the tie lines determined experimentally by Markham (1960). Unlike the experimental diagrams, Figure 19 does not necessarily portray isothermal conditions, and so the apparently conflicting tie lines drawn from sylvanite to both krennerite and calaverite represent associations probably formed at different temperatures and preserved in the ores.

INTERPRETATIONS OF THE BOULDER COUNTY ORES

Lack of Melting Phenomena

Temperatures of the liquidus in the system Au-Ag-Te (see Markham, 1960; Cabri, 1965b) are generally above the range that would seem geologically reasonable for the deposition of the Boulder County ores, but are not so high that melting phenomena can be ruled out *a priori*. As noted by Cabri (1965b), solution of water or other volatiles in liquid phases of the system

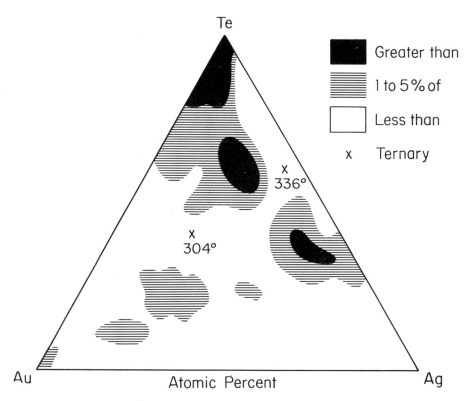

Figure 19. Natural tie lines in the system Au-Ag-Te as represented by the telluride ores of Boulder County.

would depress the melting points and hence the possibility of melting phenomena cannot be overlooked.

All textures and associations seen in the ores can be explained in terms of subsolidus relations. Micrographic intergrowths seen might be described as "eutectoid," but are actually formed either by replacement (e.g., hessite replacing tetrahedrite) or by breakdown of metastable solids upon cooling (e.g., sylvanite in hessite). Some crystals of tellurium and sylvanite in open vugs have a rounded appearance that might be due to partial fusion but the evidence is inconclusive.

If present, evidence of former liquid phases is most apt to occur in ores of the lowest melting point and hence ores of bulk composition, close to either of the two ternary eutectics in the system were carefully scrutinized. Actually, ores of such composition are, for reasons unknown, exceedingly rare in the Boulder deposits; this is apparent in Figure 20 where the ternary eutectics given by Cabri (1965b) are shown along with the bulk compositions of 156 Boulder County samples. Textures involving an original liquid phase of the kind produced synthetically in petzite-calaverite-gold assemblages by Cabri (1965b) are not present in Boulder ores of comparable bulk composition.

It is concluded that melting phenomena have played no significant role in formation of the ores, at least at the sites of deposition.

Associations of Tellurium

The observed associations of native tellurium with sylvanite, calaverite, and krennerite are consistent with the experimental tie lines of both Figure 17 and Figure 18. The association tellurium-stuetzite may, however, be a metastable one. Both Cabri (1965b) and Markham (1960) show tie lines extending from stuetzite (Ag_2Te_3) to tellurium, but the work of Honea (1964) indicates that empressite (AgTe) is a distinct phase intermediate on the Ag-Te binary between tellurium and stuetzite, and is stable at temperatures below 210°C. Empressite (AgTe) has never been synthesized, but Honea finds that natural AgTe decomposes to Ag_5Te_3 plus Te at 210°C.

Relations Along the Join $AuTe_2$-$AuAgTe_4$

Minerals along this join are calaverite, krennerite, and sylvanite. Calaverite and krennerite have long been considered polymorphs of $AuTe_2$ (Borchert, 1935; Markham, 1960), but until the recent work of Cabri (1965b) krennerite defied synthesis and as a result its stability field and relations to calaverite were unclear. Cabri proposes that krennerite is a mineral compositionally distinct from calaverite and has the fairly definite formula Au_4AgTe_{10}. He synthesized krennerite at temperatures of from 380° to 290°C and found that in the range 290° to 350°C stable assemblages are calaverite-krennerite and krennerite-sylvanite. There is some question as to the lower stability limit of krennerite. Its field may persist to low temperatures or, as suggested by occasional reports of the association calaverite-sylvanite in natural ores (Markham, 1960), krennerite may give way to calaverite-sylvanite as the stable pair at low temperature. While Markham was unable to synthesize krennerite, his heating experiments with natural material suggest that krennerite may be stable at least down to 150°C. In his experiments, natural krennerite was unaffected after being held at 150°C for seven days.

Associations observed in the Boulder ores shed some light on relations in this part of the system. There is some evidence that the stability field of krennerite must descend to temperatures at least as low as 170°C. In ores of the Alpine Horn mine, deformed early sylvanite is concomitantly replaced by petzite and krennerite (Pls. 7f and 10g). The petzite appears as replacement veinlets in sylvanite, and evidently gold released from sylvanite in the vicinity of these veinlets diffused away from the veinlets and led to formation of krennerite crystals replacing sylvanite a few millimeters away. If correctly interpreted, these krennerite crystals (Pl. 7f) formed at the same time and temperature as the nearby petzite (Pl. 10g). Cabri's experiments have shown that the association petzite-sylvanite does not form

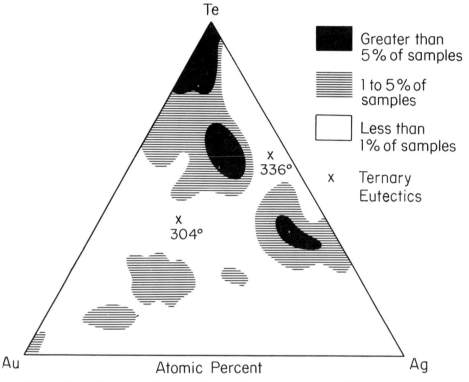

Figure 20. Bulk compositions of the Boulder County ores within the system Au-Ag-Te based on analyses of 156 polished sections.

above 170°C, and hence the krennerite in this particular case must likewise have formed below 170°C.

When krennerite is stable, the permissible associations are calaverite-krennerite and sylvanite-krennerite. These are rare in the Boulder ores because krennerite itself is uncommon. Where it is present, krennerite is associated with sylvanite, but the association calaverite-krennerite which requires a more restricted bulk composition is entirely missing.

The association calaverite-sylvanite which is stable only at temperatures below the krennerite stability field is exceedingly rare in the ores. Both calaverite and sylvanite are common but tend to occur separately in the ores. In a few specimens, sylvanite forms diffuse fringes along the edges of coarse calaverite grains (Pl. 4h) or occurs as very tiny grains in veinlets of petzite cutting calaverite (Pl. 10h). In such rare associations with calaverite, the sylvanite has only been identified by optical means and not confirmed by X-ray analysis. However, its multiple twinning, strong bireflectance, and distinctive polarization figure leave little doubt that this association is truly sylvanite-calaverite. This rare association must have formed below 290°C, the lowest established limit of the krennerite stability field.

Hessite Twinning

Practically all hessite of the Boulder veins displays twinning of the type normally ascribed to inversion from the intermediate temperature (cubic) form (Pls. 6e and f). Frueh (1959b) concludes that both natural and synthetic low temperature hessite are monoclinic and attributes the higher orthorhombic symmetry elements assigned by Rowland and Berry (1951) to twinning produced by inversion of their synthetic Ag_2Te to the low form. There are, therefore, good reasons for interpreting hessite twinning as evidence for initial deposition above the inversion point, and this criterion has been used in assigning minimum temperatures of deposition to hessite in many natural deposits (Stillwell, 1931; Markham, 1960; Baker, 1958; Callow and Worley, 1965). However, there are some uncertainties attached to the validity of this thermometer. Experimental studies have shown that the inversion takes place at 145°C in the binary Ag-Te (Kracek and others, 1966), but Cabri (1965b) suggests the possibility that it may take place at lower temperatures in the presence of gold. The present authors hesitate to apply this thermometer to the Boulder deposits, not only because the exact inversion point in the ternary system Au-Ag-Te is unknown, but also, as discussed below, the very interpretation of the twinning as an inversion feature seems incompatible with certain other phase relations.

Origin of Petzite-Hessite Intergrowths

As previously mentioned, petzite and hessite are preferentially intergrown in the Boulder County deposits and everywhere display mutual textures that strongly suggest contemporaneity. The proportions of petzite and hessite vary but are commonly about equal. The over-all impression is one of an intergrowth formed by unmixing of some originally homogeneous phase.

Cabri's (1965b) discovery of the metastable "x" phase provides the logical phase from which these petzite-hessite intergrowths were probably derived. This phase is stable over a wide compositional range along the hessite-petzite join at temperatures above 50° ± 20°C and at lower temperatures rapidly breaks down to hessite and petzite.

It is noteworthy that *all* of the hessite in the Boulder County intergrowths shows the type of twinning usually attributed to inversion from the intermediate temperature form. This is difficult to reconcile with the fact that the "x" phase breaks down directly to petzite and *low* temperature hessite; thus any hessite derived in this way should not possess true inversion twinning. It is therefore suggested that hessite derived from the "x" phase has acquired its twinning through some mechanism other than inversion, perhaps due to stresses produced in the breakdown of "x" phase. This would not preclude the occurrence of true inversion twinning in any hessite that coexisted with the "x" phase at temperatures above 50° ± 20°C, but does require a different origin for twinning in that portion of the hessite derived from the "x" phase. No differences in the patterns of hessite twinning have as yet been recognized which might serve to distinguish inverted hessite from low temperature hessite derived from the "x" phase. The writers have attempted to modify the hessite twinning by variations of sample mounting and polishing procedures, but these tests only confirm that the twinning seen is inherent in the natural hessite.

One alternative, but much less satisfactory, explanation of the petzite-hessite intergrowths would require very special conditions in order to avoid development of the "x" phase and the low temperature hessite it would produce upon cooling. This explanation calls for the replacement of previously inverted hessite by late petzite. This would allow for the presence of true inversion twinning in all of the hessite, but would also require formation of the petzite (and all other late minerals found replacing the petzite-hessite intergrowths such as altaite, tetradymite, and native gold) at temperatures below the lower stability limit of the "x" phase (i.e., $50° \pm 20°C$). Such temperatures seem excessively low even for minerals very late in the hypogene sequence. This explanation also places unusual restrictions on bulk compositions of the ore fluids just to avoid the extensive stability field of the "x" phase. Intermediate temperature hessite would first have to form without any gold-bearing "x" phase and then, after inversion of this hessite, gold-bearing petzite would have to form without any additional hessite. Finally, such an origin would account neither for the apparent contemporaneity nor the preferential association of the petzite and hessite.

In summary, the writers propose that the petzite-hessite intergrowths common in the Boulder County deposits were formed by unmixing of the "x" phase. The breakdown of this phase took place below 70°C and may have been a very late event in the post-depositional history of the ores. This explanation calls for some mechanism other than inversion to explain the twinning in hessite derived from the "x" phase and raises a new question as to the reliability of hessite twinning as a geologic thermometer.

Origin of Hessite-Sylvanite Intergrowths

Cabri's (1965b) experimental work has also revealed the probable origin of micrographic hessite-sylvanite intergrowths of the type found in Boulder County and other important telluride localities. As previously described, the sylvanite in these intergrowths occurs in variable amounts subordinate to hessite and takes the form of minute, oriented graphic blebs (Pl. 7a) or larger inclusions with a shredded or ragged appearance (Pl. 6h). The intergrowths are commonly accompanied by petzite *or* stuetzite which appear as distinct phases surrounded by, or adjacent to, the intergrown hessite-sylvanite. Intergrowths from the Cash mine contain areas of hessite free of inclusions (Pl. 7c) as well as minute grains of stuetzite that are attached to the micrographic sylvanite (Pl. 7b).

The textures and associations seen in these intergrowths suggest that they are produced by cooling of assemblages that originally contained the metastable gamma phase or metastable gold-rich stuetzites. By superposing the ternary diagrams of Cabri (1965b) and Markham (1960), it is possible to relate the initial and end products of cooling for any bulk composition that would involve these unstable materials. It can be seen, for example, that the most gold-rich stuetzite stable at 290°C happens to lie on the low temperature hessite-sylvanite join and hence stuetzite of this composition should, upon cooling, form an intergrowth consisting entirely of hessite and sylvanite. A variety of reactions involving less fortuitous bulk compositions might take place and result in the production of hessite plus sylvanite accompanied by either stuetzite or petzite. Based on comparison of Figures 17 and 18, the following combinations of initial and cooled assemblages would pertain to different bulk compositions and would account for intergrowths seen in the Boulder ores.

Initial Assemblage	Compositionally Equivalent Cooled Assemblage
Stuetzite + Gamma phase	Sylvanite + Hessite + Stuetzite
Stuetzite + Gamma phase + Sylvanite	Sylvanite + Hessite + Petzite
Gamma phase + "x" phase	Sylvanite + Hessite + Petzite

Knowing only the initial equilibrium assemblages and the stable (or metastable?) assemblages they would ultimately produce upon cooling, it is not possible to specify the intermediate steps involved in production of the hessite-sylvanite intergrowths. In other words, the precise manner in which the tie lines shift in going from Cabri's diagram at 290°C (Fig. 18) to the low temperature relations indicated by Markham's work (Fig. 17) is unknown. Some of the reactions are probably complex involving first a breakdown of the metastable gamma phase to form some "x" phase and then, at lower temperatures, the formation of hessite and petzite seen in contact in some of the intergrowths. At this point, one can only state that the textures observed in polished section suggest this sort of complex reaction origin and that the associations developed in the intergrowths are precisely those that should theoretically form on cooling of known metastable phases in this system.

Origin of Certain Late Tellurides and Native Gold

Certain Au-Ag-Te minerals occur very commonly as rims, veinlets, and pseudomorphs that *apparently* replaced older host minerals of the same chemical system. In most cases, it is impossible to deduce from microscopic evidence alone whether these minerals are exsolution products or whether they formed by attack of late vein fluids on older hosts. The problem is further complicated by the fact that the original textures, whether initially of exsolution or replacement origin, may have been modified by thermal reactions as the ores cooled. For the gold-silver tellurides it is fortunate to have the two types of phase diagrams produced by Cabri (1965b) and Markham (1960) which, when correlated with microscopic observations, allow more meaningful interpretations than are normally possible. All of the textures considered here involve some host mineral or minerals that appear to be early and a guest mineral or minerals that appear to be late; for brevity, the guest minerals will be referred to as veinlets, the form they most commonly adopt, but it should be kept in mind that these minerals also occur as rims, pseudomorphs, and irregular inclusions.

The discussion can be simplified at the start by citing the evidence indicating a replacement rather than an exsolution origin for the features to be considered. Although the textural evidence in most samples is equivocal, in some it strongly suggests a replacement origin. For example, veinlets of petzite so commonly seen transecting early sylvanite (Pl. 9b) are not restricted to this host, but rather, in some samples, extend continuously out into adjacent sulfides and sulfosalts (Pl. 10b). Also, most of the late minerals to be discussed below, including hessite, petzite, and gold, form veinlets in a great variety of sulfides, sulfosalts, and tellurides and not just the early gold-silver tellurides from which they might more reasonably be expected to exsolve. The formation of abundant pseudomorphs by some of the late minerals is also difficult to explain through exsolution. In some samples (Pls. 6a and 10c), late petzite intergrown with altaite or coloradoite (all cubic minerals) forms partial to complete pseudomorphs after bladed crystals of sylvanite or calaverite, and is too abundant to explain by exsolution from a host, little or none of which has survived in the ores. These textural indications of replacement are supported by more conclusive experimental phase relations that indicate extremely limited mutual solubility among the early and late minerals. Reference to Cabri's (1965b) isothermal sections at 356°, 335°, and 290°C shows, for example, very limited solubility along the joins calaverite-gold, calaverite-petzite, hessite-gold, and others which are precisely the tie lines between the host-guest associations considered below. Veinlets of petzite or hessite in sylvanite do present a problem from the experimental standpoint in that their associations are stable only below 170°C and their *equilibrium* relations have not been established. However, there are strong indications of very limited mutual solubility between hessite and sylvanite, and between petzite and sylvanite within the low temperature range of stability of those associations. Markham's work (1960, Fig. 17) shows no solubility along the tie lines sylvanite-petzite and sylvanite-hessite, but it is not possible to say whether these relations pertain to the entire range from room temperature up to the established maximum temperatures for these associations (i.e., 170°C) or whether they are equilibrium relations for room temperature. However, Cabri's study (1965b) has established the general pattern of solid solution in the system Au-Ag-Te in which there may be extensive substitution of gold and silver but in which the phases of the system tolerate little variation in their atomic proportions of tellurium. On this basis, it seems most unlikely that sylvanite could dissolve appreciable quantities of either petzite or hessite, phases of much lower tellurium content.

The textural and experimental evidence is therefore compatible and points to a replacement rather than an exsolution origin of the various veinlets, and these veinlets can now be discussed with other genetic problems in mind. The veinlets can be placed in one of three categories depending upon the thermal (cooling) modifications of either the wall or veinlet assemblages.

Simple Replacement Veinlets with No Thermal Modifications. Included here are veinlets of petzite in calaverite, sylvanite, or krennerite; of hessite in sylvanite; of gold in hessite or petzite; and, of intergrown sylvanite and petzite in calaverite, sylvanite, or krennerite. These veinlets have had a simple history and probably look today as they did when initially formed by replacement. There is no textural evidence nor any indication from the associations that the metastable "x" or gamma phases existed originally in either the walls or the veinlets proper (see Pls. 13d, e, f, and g). Therefore, there is no reason to suspect that the age relations or sequence of events is any different from or more complicated than the simple veining relationship suggests. Those veinlets involving contact of sylvanite with hessite or with petzite either within the veinlets or between the walls and the veinlets must have formed below 170°C according to the work of Cabri (1965b). Veinlets of petzite-sylvanite in calaverite are thought to have formed below 290°C (i.e., the lowest established temperature limit of the krennerite stability field) in order to allow sylvanite to form in contact with calaverite.

Replacement Veinlets in Which the Veinlet Minerals are Thermally Modified. These are more complex, segmented veinlets in which metastable gamma phase, "x" phase, or gold-rich stuetzites were very probably present as one of the original replacing phases and broke down upon cooling to form texturally complex veinlet assemblages. Included here are veinlets of intergrown hessite and petzite cutting sylvanite, or of micrographic hessite-sylvanite intergrowths (with some stuetzite or petzite) also cutting early sylvanite. The histories of the veinlet materials apparently involve all of the cooling reactions and textural complications previously discussed under "hessite-petzite" and "hessite-sylvanite intergrowths."

Replacement Veinlets in Which the Walls are Thermally Modified. It is fortunate that only one example arises in this category because the observed textures are quite deceiving as to the relative ages of the minerals involved. The example is that of veinlets of native gold crosscutting mutual hessite-petzite intergrowths. The phase relations leave two possibilities here: first, that the gold originally replaced "x" phase at some temperature above 50° ± 20°C and the "x" phase then formed hessite plus petzite on cooling; or, second, that the gold itself formed below 70°C and replaced the hessite-petzite directly. Of the two possibilities, the second is here rejected as requiring unreasonably low temperatures for initial deposition of the hypogene gold. One is then left with the interesting situation of native gold as antecedent replacement veinlets cutting hessite and petzite that were derived from cooling of the metastable "x" phase and are actually younger than the gold itself.

This type of analysis cannot be extended to late rims and veinlets of altaite, coloradoite, melonite, and tetradymite that are associated with native tellurium, or the early, coarse-grained gold-silver tellurides. On the basis of textural evidence, the writers believe that these are also replacement features formed in much the same way as the Au-Ag-Te veinlets discussed above, but corroborating phase relations are unknown. There has been some work on the binary system Ni-Te (Westrum and Machol, 1958) which shows negligible mutual solubility between melonite and tellurium at elevated temperatures confirming that the aggregates of fine-grained melonite commonly bordering native tellurium (Pl. 7g) or found in veinlets of petzite (Pl. 9b) or hessite (Pl. 13d) are of replacement origin.

DEGREE OF EQUILIBRIUM

Initial Deposition

The degree of local equilibrium maintained during initial deposition of the tellurides cannot be judged simply on the basis of minerals now seen in contact. While some of the present assemblages may have survived cooling without change, others have obviously readjusted and the original assemblages must be "reconstructed" before any meaningful comparison with experimental equilibrium diagrams can be made. Judgments of initial equilibrium are further complicated by the fact that initial deposition may have continued at temperatures below those for which phase equilibria have been firmly established. This is evidenced by such features as replacement veinlets of petzite *alone* or hessite *alone* (Pls. 13d, e, f, and g) in sylvanite which show no evidence of once having contained either the "x" or gamma phases and which, according to the work of Cabri (1965b), must have formed below 170°C to allow petzite or hessite to develop in direct contact with sylvanite. Although the phase relations determined by Markham (1960) correspond closely to natural associations in telluride ores from Boulder and other famous localities and might be assumed to represent equilibrium relations at room temperature, their stability has not been experimentally verified. While Markham's results are useful in predicting end products of cooling, they are inconclusive as a measure of equilibrium attained either during or after initial deposition of the telluride ores.

If we arbitrarily assume that tie lines drawn by Markham (Fig. 17) are the associations *stable* at room temperature and that these associations begin to take over upon breakdown of the gamma phase (below 170°C) and are completely established upon breakdown of the "x" phase (below 70°C), then some further speculation regarding equilibrium is possible. On this basis, it would appear that local equilibrium of the type discussed by Thompson (1958) was closely maintained on a microscopic scale during initial deposition. As the tellurides formed either by direct precipitation in open space or by replacement of one another, they did so in such a way that only a limited number of compatible phases developed in direct contact. Those assemblages that show no signs of thermal modification are precisely those synthesized by Cabri (1965b) or correspond to associations of Markham (1960). The remaining assemblages that now contain micrographic hessite-sylvanite or mutual hessite-petzite intergrowths, suggesting original "x" or gamma phases, can be explained by cooling readjustments of equilibrium assemblages of the same bulk composition on the 290° isotherm (Cabri, 1965b; Fig. 18). In both cases, the ores which have suffered no change in cooling and those in which the initial assemblages must be reconstructed, the initial assemblages contain no more than three phases of the Au-Ag-Te system in direct contact and, in this regard, meet the requirements of the mineralogical Phase Rule (i.e., number of phases $\leqq$ number of components).

This notion of equilibrium does not extend beyond the dimensions of assemblages defined by mutual contact. Roughly 10 percent of the polished sections examined contain four or more Au-Ag-Te minerals though not in mutual contact. Very clearly, still larger volumes of ore were not in internal equilibrium. The bulk composition of ore varies greatly over short distances and, as a result, individual veins may carry a large number of mutually incompatible minerals.

Cooling Stages

The experimental work of Cabri (1965b) has shown that the most rapid diffusion and reaction rates in the system Au-Ag-Te occur in assemblages of comparatively low tellurium and high silver content which contain the "x" and gamma phases at elevated temperatures, and hessite or petzite at room temperature. It is not at all strange, therefore, that ores of comparable bulk composition would show the greatest evidence of thermal readjustment nor that such ores would contain assemblages corresponding more closely to unquenched synthetic assemblages (Markham, 1960) than to assemblages in equilibrium at high temperature (Cabri, 1965b). In the Boulder County deposits, ores containing intergrown sylvanite, hessite, and petzite are abundant and match perfectly the experimental tie lines drawn by Markham. If these associations are truly stable at low temperature, then these ores maintained equilibrium throughout the cooling process.

It is more difficult to interpret ores representing other parts of the ternary system characterized by slower reaction rates. For example, the Boulder County ores containing native tellurium show no textural evidence of thermal reactions, and so might be interpreted either as assemblages stable at both elevated and room temperatures or as metastable assemblages that simply survived cooling without change. Similar uncertainties apply to unquenched synthetic assemblages which may still contain metastable high temperature associations. There is some positive evidence that equilibrium was not maintained in cooling of the Boulder ores. The authors have already mentioned the possibility based on Honea's (1964) experiments that the association tellurium-stuetzite (Ag_5Te_3) is a metastable one that should have formed empressite (AgTe) upon cooling. Also, the occurrence of sylvanite in contact with calaverite reported here and elsewhere (Markham, 1960) suggests a termination of the krennerite stability field at some temperature below 290°C, and hence that krennerite is metastably preserved.

Supergene Alteration

EXTENT

The telluride ores have not been deeply weathered and in many places primary ore was found within several feet of the ground surface. Complete oxidation typically extends to depths of from 5 to 60 feet and partial oxidation normally plays out within 150 feet of the present surface. The lower limit of oxidation coincides with the present water table which generally occurs at depths of from 50 to 150 feet throughout the area of telluride mineralization. Wilkerson (1939) mentions the occurrence of supergene gold at depths up to 140 feet in some of the larger mines of the Magnolia district where signs of oxidation very locally extended to 200 feet. The deepest occurrence of supergene gold noted in the present study was 150 feet below the surface in the Emancipation mine of the Gold Hill district where wire gold occurs in limonite. The depth of oxidation depends largely upon the openness and permeability of the vein structures and is, in general, greatest under flat areas that are remnants of old erosion surfaces (Lovering and Goddard, 1950).

Evidently, there was little difference in the character and depth of oxidation in the Jamestown, Gold Hill, and Magnolia districts, but oxidized ore formed a much larger proportion of total production in the Magnolia district. Many of the mines of the Jamestown and Gold Hill districts penetrated deeply below the limits of oxidation, some reaching depths of more than 1000 feet, whereas the majority of mines in the Magnolia district were shallow and extended to average depths of less than 200 feet (Fig. 12). This difference in depths of mining may be responsible in part for high gold:silver ratios that characterize total production from the Magnolia district.

The ground that lies beneath the thin surface shell of strong oxidation and leaching, and above the present water table, is characterized by rather spotty accumulations of iron and tellurium oxides that appear as irregular stains or are concentrated along the vein structures or crosscutting fractures. A large portion of the ore extracted from such ground consisted of primary tellurides mixed with varied amounts of limonite, tellurite, and jarosite. There is no extensive or even well-defined zone of enrichment at the base of this zone and residual enrichment of gold was very slight. Although visible gold is relatively abundant and more conspicuous in the oxide zone, the assay records suggest that total gold content was not significantly increased by surface processes.

SUMMARY OF ALTERATION PRODUCTS

General Remarks

The principal oxidation products identified in this study or previously reported were described separately under "Descriptive Mineralogy," and so the purpose in this section is to consider their genetic relationships and relative importance in the near-surface ores. Samples of typical oxidized ores removed from the old shallow workings are today difficult to obtain, and so the following discussion is based in part on the sparse early literature describing the vein outcrops and ore immediately below.

Native Gold

Extremely fine-grained native gold is released from the tellurides upon weathering and tends to remain in place as silver and tellurium are leached. In polished sections, minute particles of gold in iron or tellurium oxides can be seen in all stages of replacement of primary gold-bearing tellurides (Pls. 2a, 9c, and 12d). Disseminations of gold in limonite (goethite) were particularly common in the outcrops and comprise the miners' "rusty gold."

As pointed out by Lovering and Goddard (1950, pp. 99, 240), the lack of important placer deposits in Boulder County relates to the fine-grained character of gold derived from the telluride lodes. This material was readily dispersed in surface runoff and did not accumulate in workable stream concentrations. The only good placers found were close to veins like the Logan (Gold Hill District) in which hypogene gold was both coarse and abundant.

There is no evidence to indicate extensive solution and chemical transport of gold under supergene conditions. The assay records suggest that gold was neither added to nor chemically removed from the shallow oxide shell. Previously described films of gold found along irregular fractures in near-surface ores of the Colorado vein may be an exception, but these could also be of late hypogene origin and are certainly unusual features. The development of supergene wire gold such as that found in ores of the Red Cloud mine requires chemical migration of gold over very small distances but, compared to other metals in the veins, gold was essentially stationary.

Silver Minerals

Unlike gold, silver was extensively leached close to the surface and apparently only very minor quantities were reprecipitated at depth. Native silver was common in some of the outcrops and was accompanied by rare coatings of cerargyrite, embolite, and iodyrite; more significant quantities

of silver were carried off in ground waters. Small amounts of argentite replace galena in ores just below the water table and represent normal but
unimportant reprecipitation of silver beneath the oxide zone. Supergene
argentite (actually acanthite) makes up the "tellurium grease" that coats
the telluride ore in many of the underground workings.

Almost all of the hessite that occurs in the Boulder veins seems to be
hypogene, but minor amounts of fine-grained, supergene hessite are associated with tellurite, paratellurite, and rickardite, and replace hypogene
coloradoite in ore from the Last Chance mine. Such hessite was apparently
rare and of little economic importance.

Tellurium Oxides

Tellurium is highly mobile in the oxide zone and has been extensively
or completely leached from the vein outcrops. Both tellurite and paratellurite occur in partially oxidized ore but are not visible in the outcrops.
Tellurite has been X-rayed in several oxide samples but paratellurite in
only one; unfortunately, therefore, the relative abundance of these two
minerals in the oxide zone is unknown.

Supergene Native Tellurium

Nearly all the native tellurium is primary as evidenced by its association
with other hypogene minerals over the full vertical range of exposed ore
and by the fact that it is older than many of the hypogene tellurides with
which it is intergrown. Very minor amounts of native tellurium are possibly
supergene. Particularly suspect are slender prisms of tellurium associated
with tellurium oxide in shallow workings of the John Jay mine. In this
occurrence, tellurium is not intergrown with tellurides as it normally is.
There is also the possibility that the rare sponge variety of native tellurium
described earlier is supergene because it invariably appears as a late mineral
replacing tellurides. However, this tellurium is not associated with tellurium
oxide and occurs in ores that show no other signs of possible oxidation. It
may therefore be a product of late hypogene leaching.

Tellurites and Tellurates

The most serious gap in present knowledge of the oxide zone pertains
to the original occurrences of tellurites. A variety of tellurites may have
been present in ores mined from the shallow workings, but a careful search
has failed to reveal these in the present collection. Several samples suspected
to contain iron tellurites were X-rayed, but all materials tested proved to be
either the mineral tellurite, or jarosite. As previously noted, questionable
tellurites "magnolite" and "ferrotellurite" first reported in Boulder County
have been dropped from the record for want of definitive descriptions and
subsequent confirmations. The bismuth tellurate, montanite, was tentatively
identified by Genth (1877) in ores of the Cold Spring-Red Cloud mine, but
this occurrence lacks confirmation.

Mercury Minerals

Oxidation of the mercury telluride, coloradoite, produces some native mercury or, in the presence of silver, some Ag-amalgam. Rare drops of mercury can be found in partially oxidized ores if the time is taken to pound on hundreds of dump samples. Nowhere are the known occurrences of supergene mercury of any economic interest. Minor amounts of calomel also formed as oxidation products of coloradoite.

Copper Minerals

The copper tellurides, rickardite and weissite (?), are rare supergene alteration products. Primary vulcanite-rickardite intergrowths of the kind described by Cameron and Threadgold (1961) are altogether lacking. Traces of azurite and malachite appear in association with oxidized tetrahedrite in some ores while small amounts of supergene chalcocite replace pyrite and galena in some sulfide-rich telluride ores.

Iron Minerals

Iron released on oxidation of abundant fine-grained pyrite and marcasite of the telluride veins is partially leached from the oxide zone. The high iron content of waters draining this zone is evidenced by precipitates of transported jarosite or limonite along fractures and in pools of water standing in mine workings. Much iron lingers as goethite in the outcrops or as jarosite in partially oxidized ores beneath the outcrops. Interesting zonal relationships of goethite and jarosite previously described will be explained in later pages. The writers have not X-rayed a great many limonite samples, but several tested were goethite, and in view of previous studies (Tunell, *in* Locke, 1926, pp. 104-105; Holser, 1953; Kelly, 1958) of gossan limonites, goethite is probably the chief ferric oxide present in the oxidized ores. Occasional red stains of hematite are seen in the outcrops but these are greatly subordinate to goethite.

Supergene Gangue Minerals

A number of gangue minerals form late precipitates in cavities in partially oxidized ores and were probably formed by surface waters. Gypsum is especially common, and appears as small prisms and hairlike crystals in vugs or dusted over free surfaces in partially oxidized ores. These were probably formed from ground waters saturated by calcium from the gangue and country rock and abundant sulfate ion derived from the oxidizing ore. Such gypsum is distinctly different from the coarse selenite found at depths of 900 feet in the Ingram mine which may be a late hypogene precipitate. Some calcite is definitely supergene and occurs as minute reddish (hematitic) crystals in leached and iron-stained vugs. Films and sparse vug fillings of late opal may also be supergene. In rare cases (e.g., Horsefal vein) films of opal are concentrated on the downward sides of primary sulfides suggesting a gravitational control and perhaps precipitation from descending ground waters. Some powdery, white illite appears in late vein openings in some of the mines. For example, in ores of the Walker-Clarke stope of the Buena mine, illite and opal form hollow casts about 1 to 2 mm long that are perched on roscoelite, fluorite, and the tellurides. Both minerals evidently coated doubly terminated crystals of some other mineral, probably calcite, that was subsequently leached. The opal forms a thin, brittle outer layer supporting the inner fragile shell of clay. The illite and opal are probably supergene, but could be of late hypogene origin.

SUPERGENE GEOCHEMISTRY

General Remarks

In the following pages, theoretical reasons are given for the geochemical behavior characteristic of the chief metals—tellurium, gold, silver, and iron—in the zone of weathering. This is best accomplished with Eh-pH or other computed equilibrium diagrams that define the stability limits of various ions and minerals in terms of certain controlling variables such as oxidation potential, acidity, and halide concentrations. The computational methods employed in constructing these diagrams are outlined in great detail by Garrels (1960) and will not be discussed at length here. Following Garrels, the authors have selected a value of 10^{-6} moles per liter as the maximum activity of any ion in equilibrium with minerals that can be considered insoluble over geologic spans of time. This choice is reflected by the position of all ion-solid boundaries in the diagrams that follow. The basic thermochemical data upon which the calculations are based are those given in the compilation by Garrels and Christ (1965) unless other sources are specified. At present, there is a regrettable lack of data for naturally occurring tellurium minerals, especially the tellurites and tellurates, and hence the equilibrium diagrams presented here are, at best, first approximations of more complex natural systems.

Behavior of Tellurium

The equilibrium Eh-pH diagram for the system Te-H$_2$O prepared by Deltombe and others (1956) explains some properties of tellurium in the oxide zone. A part of this diagram is reproduced in Figure 21 with only minor modifications. Of particular geologic interest are those fields and boundaries that fall within the shaded area of the diagram. This area represents the full range of acidities and oxidation potentials to be expected in mine waters; it is based in part upon actual field measurements (Becking and others, 1960; Sato, 1960) and upon certain reactions that set lower theoretical limits upon oxidation potentials developed in the presence of pyrite (Sato, 1960). The following reactions were chosen as theoretical lower limits of oxidation potential in the present work:

(1) $\quad 10H_2O + FeS_2 = FeO \cdot OH + 2SO_4^= + 19H^+ + 15e^-$
$$E = .374 - .075\ pH + .008\ \log(SO_4^=)$$

(2) $2H_2O + Fe^{++} = FeO \cdot OH + 3H^+ + 1e^-$
   $E = .722 - .177\,pH - .059 \log (Fe^{++})$.

In writing these reactions and computing the boundaries shown in Figure 21, the authors have used data for the mineral goethite, $FeO \cdot OH$ (Schmaltz, 1959), rather than ferric hydroxide used by Sato (1960). Toulmin and Barton's value (1964) of -38.120 kcal for the free energy of formation of pyrite was also used. All other boundaries around the area representing probable Eh-pH conditions are based on numerous Eh-pH measurements in actual mine waters (Sato, 1960; Becking and others, 1960). Although there is no sharp Eh-pH boundary that can be drawn to separate the intergradational environments of the oxide and primary ore zones, the approximate location of this transition drawn by Becking and others (1960) is shown by the heavy dashed line in Figure 21 and other Eh-pH diagrams that follow.

The extensive leaching of tellurium noted in the outcrops of the Boulder veins is readily explained by the fact that various ions predominate over solids under a variety of conditions reasonable for the oxide zone. Close to the surface, tellurium should move either as TeO_2H^+, aqueous H_2TeO_4, or $HTeO_4^-$. Under conditions of moderate oxidation potentials and acidities likely to exist in deeper parts of the oxide zone, tellurium should either migrate as $HTeO_3^-$ or be temporarily precipitated as tellurium oxide. Once having reached the water table, tellurium could be readily dispersed as $HTeO_3^-$ or $TeO_3^=$ in neutral to alkaline waters of low oxidation potential.

Minerals of this system stable within the probable Eh-pH range are paratellurite (tetragonal TeO_2) and native tellurium. No thermodynamic data are available for tellurite (orthorhombic TeO_2) and hence its stability relations are unknown. There is some microscopic evidence (Switzer and Swanson, 1960, p. 1274) that tellurite is metastable with respect to paratellurite. In the following discussion, it will be assumed that the metastability (?) field of orthorhombic TeO_2 (tellurite) corresponds more or less with the field for tetragonal TeO_2 (paratellurite) shown in the Eh-pH diagrams.

It is most interesting that native tellurium has a stability field extending into the environment of the oxide zone. Under rather restricted conditions of moderate Eh (ca. .12 to .25 volts) and pH (ca. 4.5 to 5.8), native tellurium could form stably and could also coexist with paratellurite under conditions defined by a narrow segment of the Te-TeO_2 boundary. This certainly supports the possibility that unusual tellurium crystals associated with tellurite in shallow ores of the John Jay mine are supergene. The rarity of such occurrences is a logical consequence of the restricted Eh-pH conditions under which they can form. It should be stressed here that the stability fields shown for both TeO_2 and native tellurium are maximum fields; the extent to which other minerals like the tellurites encroach upon these fields cannot now be determined.

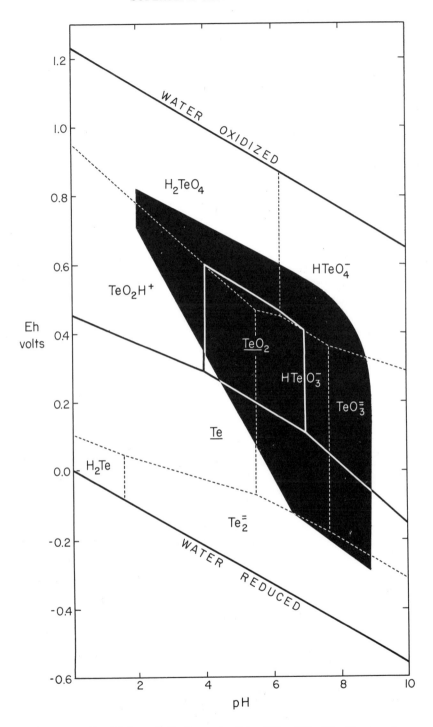

Figure 21. Equilibrium oxidation potential-acidity (Eh-pH) diagram for the system Te-H$_2$O at room temperature (from Deltombe and others, 1956).

Behavior of Gold

The very special conditions of high acidity, oxidation potential, and chloride ion activity required for the solution and transportation of gold were most recently defined by Cloke and Kelly (1964). Solubilities in excess of 10^{-6} moles per liter require an improbable combination of pH at least as low as 4.8 with oxidation potentials at least as high as .88 volts in waters abnormally rich in chloride ion. While some mine waters contain the necessary chloride concentrations, the required Eh-pH combination has not yet been recorded in actual measurements (Sato, 1960; Becking and others, 1960). The extreme character of these conditions is also illustrated by the fact that solutions of this nature would release free chlorine gas. There are some examples of gold transport in supergene zones that are well established by field evidence (Emmons, 1917). These are commonly deposits in arid or semi-arid country where there is a surface accumulation of chlorides due to evaporation and where the primary gold is accompanied by abundant pyrite as a source of acid, and manganese minerals as a source of strong oxidants. Although they contain abundant pyrite, the Boulder County ores are practically free of primary manganese minerals and there are certainly no discernible chloride concentrations near the present surface.

In Figure 22 the position and extent of the gold solubility field under very favorable conditions of high chloride (10^{-1} moles per liter) in the surface waters is indicated. Typical mine waters contain from 10^{-5} to $10^{-1.5}$ moles Cl$^-$ per liter (Hodge, 1915; Emmons, 1917), and this is a generous assumption of Cl$^-$. Within this field, gold dissolves as the auric chloride complex, AuCl$_4^-$. The remote location of the gold solubility field with respect to the range of probable Eh and pH values also indicated in Figure 22 is entirely consistent with the observed immobility of gold in almost all of the Boulder County veins. Within the range of probable conditions, no more than several parts per *billion* of gold would dissolve at any time, and so it is not surprising that the small degree of solution and migration required for development of wire and leaf gold is all that was accomplished in the many years that these veins have weathered. The theoretical relations do not prove either a supergene or hypogene origin for the unusual films of gold described in shallow ores of the Colorado vein near Gold Hill. The localization of this gold along late fractures close to the surface suggests a supergene origin, but calls for supergene conditions not met elsewhere in the telluride belt. There is no limonite or pyrolusite associated with this gold nor do the writers actually know how deep it persists below the shallow mine workings. Presumably, late acid, chloride-rich hypogene waters ascending locally may have precipitated this unusual gold.

The theoretical relations already introduced for tellurium and gold can, at this point, be combined to explain some major trends evident in the paragenesis and zonation of oxidation products. The path a-b, drawn across the area of probable conditions in Figure 22, shows the sequence of some oxidation products to be expected in the alteration of pyritic, gold telluride

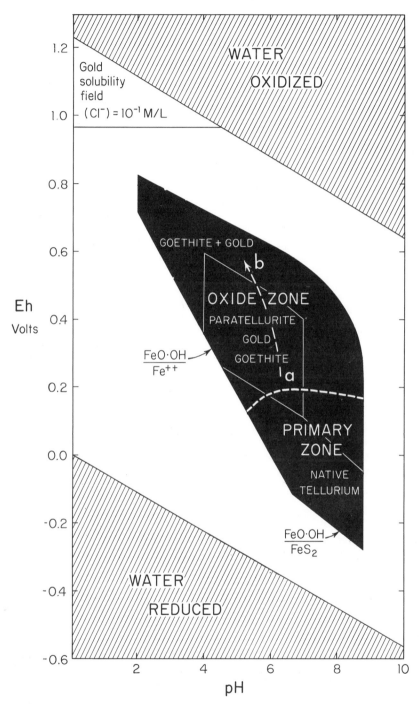

Figure 22. General Eh-pH relations in near surface telluride ores of Boulder County.

deposits. The trend is one toward assemblages stable under progressively higher acidities and oxidation potentials, and should be expressed both by the paragenetic sequence seen in individual samples and by the changes noted in tracing primary ore upward through zones of increased oxidation. Keeping in mind that both goethite and gold are insoluble throughout the shaded area of Figure 22, the theoretical sequence is (1) unaltered primary ore, (2) goethite $+ TeO_2 +$ gold, and (3) goethite $+$ gold. This is precisely the sequence noted in the Boulder County deposits where tellurite and paratellurite appear as temporary oxidation products in partially oxidized ore but in time are leached. Only goethite and gold persist through the process, and these tend to remain in the outcrops in the form of rusty gold.

Behavior of Silver

Under conditions likely to exist in the oxide zone, the behavior of silver is governed primarily by prevailing oxidation potentials and chloride activities with acidity playing a subordinate role. Theoretical stability relations among important silver species of the system Ag-Te-S-H-O-Cl are expressed in terms of oxidation potential and chloride ion activity in Figure 23. A pH of 4 was assumed in computations of equilibrium formulae for reactions involving H+, and the upper and lower limits of the diagram are the stability limits of water at this pH. For this particular diagram, the solutions are assumed to be saturated with respect to tellurium ($\Sigma Te=10^{-7}$ moles per liter) and sulfur ($\Sigma S=10^{-1}$ moles per liter) at 25°C and one atmosphere total pressure. The sizes of the stability fields shown for native tellurium, native sulfur, and tellurium oxide are in part determined by this assumption. A value of -8.7 kcal used as the free energy of formation of Ag_2Te at 25°C was estimated by extrapolation of $\Delta F°$ $_{Ag_2Te}$ determined at 250° and 300°C by Kiukkula and Wagner (1957). Free energies of formation of $AgCl_2^-$ and $AgCl_4^{\equiv}$ were computed from solubility constants determined by Berne and Leden as listed by Bjerrum and others (1958).

As in the previous Eh-pH diagrams, it is possible to outline the area within Figure 23 that represents the ranges of Eh and chloride ion activity (a $_{Cl^-}$) likely to be encountered in mine waters. This area, shaded in Figure 23, is bounded by extreme values of chloride ion concentrations revealed by analysis of mine waters (Hodge, 1915; Emmons, 1917) and by Eh limits previously discussed and shown at a pH of 4 in Figures 21 and 22.

The mobility and mineralogy of silver in the Boulder County veins are entirely consistent with the relations indicated by Figure 23. The reported occurrences of native silver and cerargyrite might be expected from the sizable stability fields of AgCl and Ag° within the limits of probable conditions. From the early literature on these deposits, it appears that cerargyrite was quite rare in comparison to wire silver. This, along with the observed leaching of silver, suggests that oxidation of the ores was accomplished in a low chloride environment (a$_{Cl^-}<10^{-3.8}$ moles per liter).Evidently, near-surface accumulations of chlorides are associated with high rates of evap-

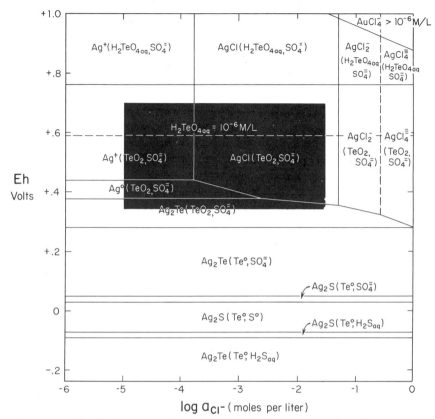

Figure 23. Equilibrium oxidation potential-chloride ion activity diagram for the system Ag-Te-S-H-O-Cl at room temperature.

oration in arid or semi-arid climates and had no opportunity to form under conditions prevailing since exposure of the Boulder deposits to the weathering environment.

Extensive leaching of silver from outcrops of the telluride veins can be explained as a result of the large field of predominance of Ag^+ under reasonable Eh-Cl^- conditions. Within this field, silver solubilities should range from $10^{-4.8}$ to 10^{-6} moles per liter. It is significant that the known silver chloride complex ions require chloride ion activities somewhat higher than those of normal mine waters. From this it is concluded that silver leached from the Boulder veins must have moved as Ag^+ rather than as $AgCl_2$ or $AgCl_4$. In the upper right corner of Figure 23 the field of gold solubility is shown as it exists at a pH of 4. Here, again, this field is well removed from the range of probable conditions and clearly separate from the field of Ag^+. It is not surprising therefore that gold should persist in the outcrops as silver is leached.

A stability field for Ag_2Te (plus TeO_2) appears just below the $AgCl$ and Ag° fields and within feasible limits of Eh suggesting conditions for the formation of supergene hessite seen in partially oxidized ores of the Last Chance mine. The full extent of the hessite stability field, its dependence upon pH, and its relations to argentite (acanthite) are most clearly illustrated by the Eh-pH diagram shown in Figure 24. This diagram is similar to one prepared for the system Ag-S-O-H-Cl by Szekely (in Schmitt, 1962), but incorporates changes that result from saturating this system with tellurium. Hessite is stable over a wide range of conditions to be expected in both the oxide and primary ore zone. This suggests that hessite might play a role in some telluride deposits analogous to supergene argentite in sulfide deposits and that silver enrichment in the form of hessite should be anticipated below the oxide zone in deeply weathered deposits. Economically significant enrichment is lacking in the Boulder deposits where the oxidation is very shallow. Evidently erosion kept pace with oxidation of these ores and thereby limited the supply of Ag^+ to deeper parts of the veins where supergene hessite might otherwise have formed in abundance.

Very minor amounts of argentite occur as supergene replacements of galena and also comprise the "tellurium grease" found on some of the mine workings. As shown in Figure 24, the stability field of Ag_2S falls just below the lower limits of Eh as set by the pyrite-goethite equilibrium boundary. This is generally true whether or not the mine waters contain tellurium and suggests that actual mine water potentials must reach values below this theoretical limit. The minor argentite formed by sulfide replacement beneath the water table in some of the telluride veins probably developed under weakly acid to weakly alkaline, reducing conditions as required by part of the Ag_2S field shown in Figure 24. It is more difficult to account for the formation of argentite in tellurium grease exposed to air in the moist mine workings. Under these conditions, oxidation potentials must be well above those of the argentite stability field even at very high acidities. This suggests that this argentite formed as a metastable precipitate on the mine workings.

Jarosite-Goethite Relations

The behavior of iron in the oxide zone is entirely normal for pyrite-rich deposits with some enrichment by limonite accumulation at the surface and the drainage of ferrous ion in acid sulfate waters descending through the veins. However, the conspicuous zonal relationships of supergene jarosite and limonite are of more than routine interest. Furbish (1963) reviews the general problem of jarosite stability, and describes an intergradation of jarosite and goethite along fractures in a molybdenite prospect in Halifax County, North Carolina, which seems very similar to the type of zoning observed in some of the Boulder specimens. He attributes the gradation from jarosite to goethite to dilution of waters away from a source of jarosite components (a pyritic sulfide deposit in a granite host). Increased dilution in the direction of transport causes a rapid increase in pH and drop in ion concentrations (K^+ and $SO_4^=$) thus favoring goethite at greater distances

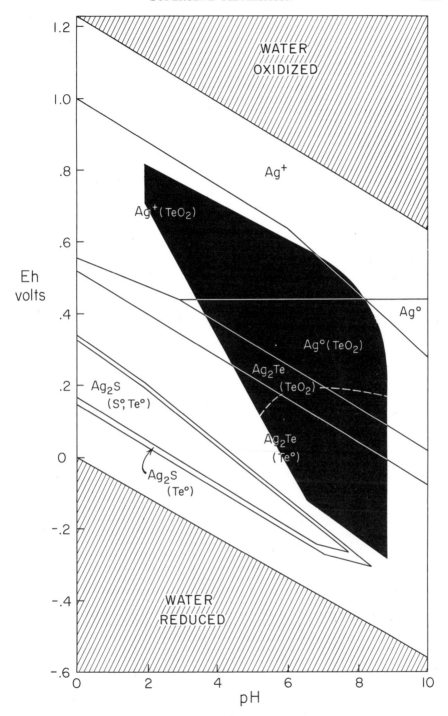

Figure 24. Equilibrium oxidation potential-acidity (Eh-pH) diagram for the system Ag-Te-S-H-O-Cl at room temperature.

from the source. His explanation is entirely compatible with the interpretation that follows, but the Boulder County ores show somewhat different types of jarosite-goethite zoning.

From experimental studies of the system Fe_2O_3-SO_3-H_2O (Posnjak and Merwin, 1922) it is known that hydrous basic ferric sulfates similar to jarosite will form from ferric sulfate solutions if these are sufficiently concentrated and the ratio Fe^{+++}/SO_4 is high. If the solutions are too dilute, goethite forms. If the solutions are concentrated but the ratio Fe^{+++}/SO_4 is too low, then normal (acid) ferric sulfates form. These early experiments therefore suggest the factors that control jarosite-goethite relations, but unfortunately potassium was not included in these experiments and there has been no recent work on the system K-Fe-S-H_2O which clarifies the present jarosite-goethite relations. However, some inferences can be drawn from reactions involving these minerals. In the oxide zone, both minerals will dissolve according to the following reactions:

Jarosite
$$KFe_3(SO_4)_2(OH)_6 + 6H^+ + 3e^- = 6H_2O + K^+ + 2SO_4 + 3Fe^{++}$$

Goethite
$$FeO \cdot OH + 3H^+ + 1e^- = 2H_2O + Fe^{++}$$

It is evident that solution of both minerals is favored by relatively high acidity and low oxidation potentials. Increasing amounts of K^+ should inhibit the solution of jarosite but have no *direct* effect on the solution of goethite. Finally these reactions may be combined or the reaction may be written directly to express the equilibrium between goethite and jarosite as follows:

Jarosite　　　　　　*Goethite*
$$KFe_3(SO_4)_2(OH)_6 = 3FeO \cdot OH + K^+ + 2SO_4 + 3H^+$$

From this reaction it is clear that the formation of jarosite as opposed to goethite is favored by high acidity, high a $SO_4 =$, and high a K^+, in that order of importance. The actual equilibrium constant for this reaction is unknown nor can one be computed because the free energy of formation of jarosite is also unknown. However, even the reaction as written will permit some speculation regarding reasons for the curious goethite-jarosite relations in the Boulder ores.

The tendency of jarosite to predominate over goethite in or closer to veinlets of pyritic horn quartz is a consequence of expected pH and SO_4 gradients. Maximum acidities and sulfate concentrations should exist at the site of oxidizing pyrite-marcasite within the veinlets and would decline toward the adjacent wall rocks. Provided that the concentrations of H^+ and SO_4 were sufficiently high, one would predict on theoretical grounds that a central zone of jarosite would develop and be flanked outward by zones of goethite formed at lower acidities and sulfate concentrations. This is precisely the relationship found in many of the oxidized telluride ores.

Potassium ion gradients might enhance or interfere with this effect depending upon their magnitude and orientation with respect to the veinlets, but potassium ion would be much less effective than H^+ or SO_4^- which enter the jarosite-goethite equilibrium constant with higher exponents. In the Boulder ores, potassium is provided by sericite present in both the veinlets and in greater abundance in the altered wall rocks. Apparently there was sufficient K^+ present wherever high concentrations of H^+ and SO_4^- called for the precipitation of jarosite, but K^+ does not appear to have been a dominant control of jarosite distribution.

The common occurrence of both jarosite and goethite as transported minerals along fractures and within vugs is consistent with the fact that both these minerals will dissolve if the pH and oxidation potentials are sufficiently low. Reprecipitation could be triggered by neutralization, oxidation, or simply by evaporation of the transporting solutions, and in any given sample it might be difficult to determine which mechanisms were involved. Samples in which vugs are lined by goethite and a distinct inner and younger layer of jarosite display the sequence to be expected from evaporation of acid sulfate waters containing potassium ion. Early evaporation would first produce saturation with respect to goethite and continued evaporation would build up SO_4^- and K^+ concentrations to the levels required for jarosite. In many places jarosite and goethite are distinctly segregated, but irregularly distributed along fractures where place to place variations of K^+, SO_4^- and H^+ were probably also irregular.

The Ore-Forming Environment

DEPTHS AND PRESSURES OF ORE DEPOSITION

According to Lovering and his co-workers (Lovering, 1941; Lovering and Goddard, 1950; Lovering and Tweto, 1953), the surface existing at the time the telluride and tungsten ores were formed was the Eocene Flattop surface of Van Tuyl and Lovering (1935). If this is assumed to be correct, then depths of ore deposition relative to the Flattop surface can be estimated. Numerous remnants of this surface are preserved at elevations of from 11,000 to 12,000 feet about 8 to 15 miles west of the telluride belt, and these slope gently and with decreasing gradients to the east. In Figure 25, topographic profiles are shown in which the Flattop surface is projected eastward to obtain its approximate original elevations in the Jamestown, Gold Hill, and Magnolia mining districts. The sections are drawn along straight lines oriented approximately east-west and with a vertical exaggeration of about 3:1. Closely comparable Flattop elevations in the range 9700 to 10,400 feet are indicated for all three districts. The telluride ores which have been mined at elevations of from 5795 to 8750 feet were 2600 to 4600 feet below the levels of the reconstructed Flattop surface. Similar figures were deduced by Lovering (1941) who concludes that the ores were formed beneath a cover 2000 feet or more in thickness.

Wahlstrom (1947) has questioned Lovering's interpretation and believes that the Flattop surface was developed in late Pliocene or early Pleistocene time, and was subsequently warped and vertically displaced by faulting. According to his view, the telluride belt has been downdropped relative to the high country to the west, and the Flattop surface is now represented in the mining districts by accordant surface remnants found at elevations of from 7600 to 8600. As indicated in Figure 25, this lower surface is very well defined in the area of telluride mineralization and bears a step-like relationship to remnants of the high surface. Some of the veins (e.g., Golden Age, Jamestown district) crop out on the lower surface, and the high flat ground and summit levels throughout the telluride belt generally coincide with this surface. Van Tuyl and Lovering (1935) called this lower surface the

165

Overland Mountain peneplain and considered it Oligocene, and, therefore, younger than the Flattop surface.

If we assume that Wahlstrom's interpretation is correct, then the actual position of the surface that existed during mineralization is indeterminate save that it must have been somewhere above the Overland Mountain surface. It is still of interest to note the position of the lowest ore deposits because their depths below the Overland Mountain surface are minimum depths of formation. In the Gold Hill district, telluride ores have been mined at elevations as low as 5795 feet, about 2600 feet below the lowest probable elevations of the Overland surface in that district. Minimum depths estimated for the deepest known ores in the Jamestown and Magnolia districts are 1900 and 1700 feet respectively.

For the restricted area involved in these speculations, the present writers prefer the physiographic interpretation of Lovering at least with respect to the well-defined Flattop and Overland Mountain surfaces, and their time and spatial relationships to the telluride ores. The interfingering of these two surfaces in plan view and their separation by consistent vertical intervals throughout the area considered are normal relationships for erosion surfaces of different age, but do not appear to fit a fault pattern. Furthermore, the faults or fault zones required by Wahlstrom's hypothesis are inferred from the erosion surfaces themselves and have never been found in the field.

The authors therefore propose that the known telluride ores were formed under a rock cover 2600 to 4600 feet thick. Ore undoubtedly extended to shallower depths above the present outcrops and shows every indication of persisting below the current mining limits thus enlarging the inferred interval of mineralization. Pressures exerted on the ascending ore fluids are difficult to estimate, but probably ranged between minimum hydrostatic pressures of from 78 to 137 bars and lithostatic pressures of from 200 to 360 bars (assumed load of from 2600 to 4600 feet of Boulder Creek granite, bulk density = 2.61). Pressures may even have exceeded the lithostatic limit where the fluids encountered restrictions to flow, but we suspect from the vuggy textures of the ores and the multitude of interlacing and well-connected fractures that comprised the channelways that such buildups of pressure were uncommon.

The arsenopyrite barometer (Clark, 1960a and b) provides an interesting check on these physiographic estimates of depth. By experimental studies of the system Fe-As-S, Clark has shown that the composition of arsenopyrite formed in equilibrium with pyrite can be used to determine approximate depositional confining pressures provided some independent estimate can be made of the pyrite-arsenopyrite depositional temperatures. Use of the barometer also requires the assumption that the arsenopyrite has retained its original composition during all events subsequent to its deposition. As previously described, arsenopyrite is intergrown with pyrite and marcasite

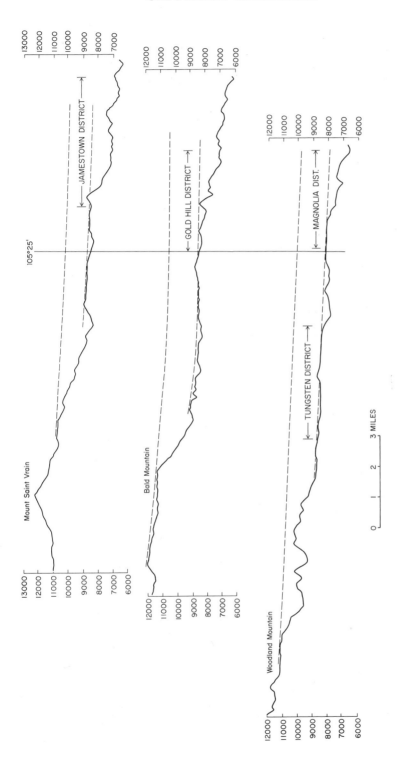

Figure 25. Topographic profiles showing reconstructed positions of the Flattop erosion surface in the Jamestown, Gold Hill, and Magnolia districts.

in horn quartz of the Archer mine, and appears to have formed contemporaneously with the pyrite. The d_{131} spacing of the arsenopyrite is 1.6307 ± .0002 Å as measured against the d_{311} reflection of fluorite (Swanson and Tatge, 1953). According to the data of Morimoto and Clark (1961, Fig. 3), this spacing indicates a sulfur-rich arsenopyrite containing 30.7 ± 1 atomic percent arsenic. As discussed in the following section, all of the thermometers applicable to the telluride ores indicate depositional temperatures below 350°C, and most of the ore is thought to have formed below about 250°C. If 350°C is accepted as a reasonable maximum for the arsenopyrite as well, a maximum depositional confining pressure of approximately 400 bars is indicated by the d_{131}-P-T curves of Clark (1960b, p. 129). This independent estimate agrees remarkably well with the pressure range of from 78 to 360 bars determined on the basis of physiographic evidence.

TEMPERATURES OF ORE DEPOSITION

Qualitative Indications

Many features of the telluride deposits such as the fairly open and vuggy ore textures, the predominance of open filling over replacement, and the very fine grain of the pyritic horn quartz suggest their formation in a moderately shallow environment at low temperatures. The general character of the wall rock alteration is typical of that normally associated with moderate-to-low temperature veins. The presence of abundant tellurides is not in itself indicative of a specific temperature range although these minerals are evidently most abundant in the mesothermal to epithermal range of the traditional intensity scale. More indicative is the complete lack of pyrrhotite, magnetite, or lime silicates that are usually present and abundant in high temperature veins.

In addition to these qualitative guides, there are fortunately a number of temperature-dependent features and assemblages in the Boulder deposits that help to narrow down the actual range of depositional temperatures. The writers have attempted to apply a variety of these thermometers as checks upon one another and thereby establish the depositional range as firmly as possible. For all of the thermometers discussed below that require some independent estimate of confining pressure, we have assumed the extreme range of a minimum hydrostatic pressure of 78 bars at 2600 feet to a lithostatic pressure of 360 bars at 4600 feet as estimated on the basis of available physiographic evidence previously discussed.

Au-Ag-Te Phase Relations

Phase relations in this system were previously discussed at length, so their thermometric applications will be treated very briefly here. Many of Cabri's experimental findings (1965b) offer potential thermometers some of which are applicable to the Boulder ores. The melting of calaverite at 464 ± 3°C sets a maximum for this mineral and, along with the melting point of tellurium at

449.7 ± .2°C (Machol, 1959), sets a rather high depositional limit upon the tellurium mineralization. More useful are the maxima determined by Cabri for incongruent melting of krennerite at 382 ± 5°C and sylvanite at 354 ± 5°C. Cabri's isothermal sections at 356°, 335°, and 290°C show no tie lines extending from sylvanite to hessite or to petzite, and he states that samples heated at 170°C gave the same data as at 290°C (Cabri, 1965b, p. 1592). Therefore, the associations sylvanite-hessite and sylvanite-petzite that are so common in Boulder County must have formed at some temperature below 170°C. Based on Cabri's experiments, the krennerite stability field is known to persist to temperatures as low as 290°C, and this sets a maximum limit for the rare association sylvanite-calaverite. The significance of hessite twinning as a thermometer was previously questioned because (1) the inversion temperature(s) in the system Au-Ag-Te are not established (see Cabri, 1965b, Fig. 7 and p. 1585) and (2) twinning may be produced in hessite by other mechanisms and may be indistinguishable from true inversion twinning (see earlier discussion of petzite-hessite intergrowths). The lower stability limit of the "x" phase, 50° ± 20°C, places an upper temperature limit of 70°C on the petzite-hessite intergrowths.

Honea (1964) reports that natural empressite (AgTe) decomposes to stuetzite (Ag_5Te_3) plus tellurium at 210°C which places a maximum temperature of formation on the rare occurrences of this mineral in the Empress mine of the Gold Hill district.

Arsenopyrite

Based on Clark's experimental study (1960a) of the system Fe-As-S, a maximum depositional temperature of approximately 490°C can be assigned to intergrowths of contemporaneous pyrite and arsenopyrite in horn quartz of the Archer mine. The arsenopyrite-pyrite solvus was previously employed as a barometer to check the physiographic estimates of depositional confining pressures. This solvus is much less sensitive as a thermometer, but can be so applied in conjunction with the independent physiographic estimates of confining pressure. It should be stressed that the sulfur-rich composition of the Archer mine arsenopyrite and the fairly low confining pressures estimated as 78 to 360 bars require extrapolation of Clark's d_{131}-P-T curves (Clark, 1960b, p. 129) 30° beyond their plotted limits in order to make the temperature estimates. With this limitation, depositional temperatures in the range 330 to 345 are indicated for the arsenopyrite-pyrite assemblage.

Bravoite

Kullerud (1962) reports that bravoite decomposes to FeS_2 plus NiS_2 at temperatures above 137° ± 6°C. This places an upper temperature limit on bravoite in ores of the Osceola-Interocean mine, and presumably on hessite, petzite, and melonite that coat this bravoite. The intergrown petzite and hessite were very probably derived from original "x" phase which unmixed at some temperature below 70°C. Assuming descending depositional temperatures throughout the sequence, the original "x" phase would also have formed at temperatures below 137° ± 6°C in this particular occurrence.

Goethite-Hematite Relations

A very useful maximum temperature can be set on occurrences of hypogene goethite in the telluride veins and also the tungsten veins (Lovering and Tweto, 1953), which further confirms the low depositional temperatures for late tellurides indicated by such thermometers as bravoite and breakdown of the gamma phase. A pertinent equilibrium P-T curve has been computed by Garrels and Christ (1965, pp. 338-341) from experimental data of Schmaltz (1959). At pressures in the range 78 to 360 bars, the highest formational temperatures for goethite are 133° to 147°C. The hematite-goethite equilibrium is not strongly affected by pressure and thus, even if the present estimates of pressure are mistaken, a maximum of about 175°C would apply to goethite formed at pressures as high as 1000 bars.

Wall Rock Alteration Assemblages

Before turning directly to the few specific thermometers in this category, some general matters relating wall rock and vein thermometry must be considered. There is first the question of whether temperature-dependent wall rock assemblages reflect the temperatures that prevailed in the vein fluids or much lower temperatures. Lovering (1950) has shown that after a brief initial period of rapid heating of the wall rock, rock temperatures many feet from a vein become practically the same as those in the vein itself, and thenceforth mineralization proceeds in a laterally isothermal environment. In this case, wall rock assemblages having known upper temperature limits of stability should be useful in determining maximum temperatures that characterized the ore fluids. On the other hand, if the mineralization is interpreted as a short-lived or even catastrophic process, high thermal gradients might envelop the veins and the ore minerals might form at temperatures much higher than those suggested by the wall rock assemblages. Under these conditions, vein temperatures might fluctuate rapidly and the changes might never be relayed far into the adjacent wall rocks.

The factors of time and proximity to the vein are therefore critical in applying the wall rock assemblages to the thermometry of the vein fluids. Some feeling for the amount of time required to achieve the isothermal environment is provided by heating curves of the type computed by Lovering (1950, p. 250-251, 254) for a long time interval. The present authors have prepared two sets of such curves to illustrate conditions during the important early stage of rapid heat conduction and to simulate as closely as possible the situation presented by the Boulder County telluride veins. These curves, shown in Figure 26, are computed from classical formulas of linear heat conduction that are fully discussed by Ingersoll and others (1954,

p. 86-91). The curves show the theoretical temperature distribution in granite walls adjacent to a vein that is supplied from depth with hot fluids that are arbitrarily assumed to be at a constant temperature T_s of 300°C. The temperature T at any distance x (cm) from the vein after an interval of time t (sec) is given by the formula

$$T = T_s - T_o \, \Phi \, (x/2\sqrt{\alpha t})$$

where T_o is the initial temperature, α the thermal diffusivity of the granite, and Φ is the standard error function of the term enclosed by brackets (see Ingersoll and others, 1954, Appendix D). The set of dashed curves in Figure 26 is based on improbably high assumptions of diffusivity (.020) and initial temperature (100°C) in the granite, and for this reason show the minimum amounts of time that thermal gradients would persist in the walls. More realistic assumptions were made in computing the solid curves in Figure 26 ($T_0 = 80°C$, $\alpha = .012$). The small differences between these two sets of curves show that the calculated results are not seriously affected by accuracy of the assumptions that must be made. In either assumed situation, thermal gradients would become negligible within a period of 100 years. In shorter times, however, appreciable gradients would still exist, and the significance of wall rock assemblages as vein thermometers depends upon the time lapsed and the distance between the mineral assemblage and the vein proper. For example, temperatures prevailing after ten years in outer parts of the alteration along the tungsten veins, perhaps 30 feet from the veins, would be 60° lower than those of the vein fluids, while temperatures developed in the same time interval in outer parts of narrow telluride alteration, perhaps 3 feet from the veins, would be within 10° of the vein fluids.

Normally one could ignore the conditions and effects of this early heating stage, but Lovering (1941) and Lovering and Tweto (1953) contend that both the telluride and tungsten ores of Boulder County were formed in a short-lived, explosive period of mineralization. If this is true, then it would be unwise to equate the maximum wall rock and vein temperatures, especially if the explosive mineralization is thought to have lasted no more than perhaps a month or several months. If the explosive event lasted steadily for at least a year, then errors in applying the wall rock thermometers to the vein fluids would be quite small for the telluride veins and their narrow alteration envelopes, but would still be excessive for the tungsten deposits. For these, it would take almost one hundred years of continuous hydrothermal flow to achieve the isothermal condition and this seems to be a rather long period of time, even on a geological scale, to be described in terms of an explosive mineralization. However, since Lovering and Tweto (1953) do not assign any important role to temperature gradients in their theory of hydrothermal alteration, we infer that by sudden and explosive they mean a mineralization that took at least one hundred years but not much more.

The duration of the telluride mineralization is further discussed in later pages where evidence is cited to support a gradual mineralization by slowly

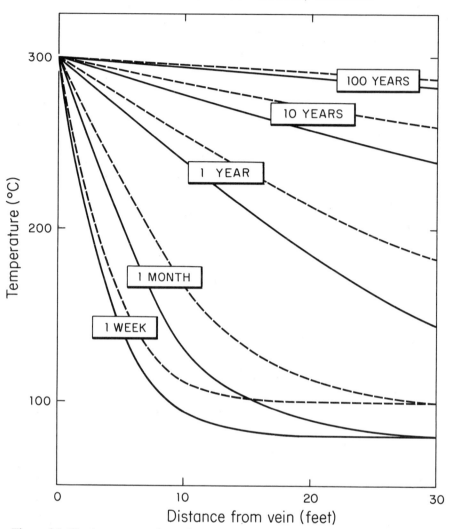

Figure 26. Heating curves of granite wall rocks exposed to vein fluids at 300°C for different time intervals up to 100 years.

moving ore fluids which might have taken thousands of years. The writers believe that the time involved in the early period of rapid heat conduction was insignificant compared to the total time of mineralization. For this reason we also feel that hydrothermal alteration assemblages occurring within several feet of the veins are significant not only in setting upper limits on wall rock temperatures, but on average temperatures of vein fluids as well.

Two wall rock associations seem to be significant in this regard. The association kaolinite-quartz, which occurs both in veinlets and in the outer zone of argillic alteration (Bonorino, 1959), probably formed at temperatures below 385°C, the lower limit of the stability field of pyrophyllite (Hemley and Jones, 1964, pp. 563-564). Fournier (*in* Roedder, 1965, p.

1392) has studied the conditions that might account for the curious kaolinite-secondary K-feldspar zone that borders some of the telluride veins (Bonorino, 1959), and concludes that metastable high silica activities depress the muscovite stability field below the limits determined by Hemley and Jones (1964) and this allows the association to develop at temperatures below about 250° to 300°C at 1000 bars pressure. It is reassuring that the temperatures suggested by these wall rock assemblages are quite reasonable when compared with almost all of the more direct vein thermometers.

Fluid Inclusion Thermometry

Most of the transparent vein minerals in the Boulder deposits are much too fine grained for satisfactory visual fluid inclusion thermometry. Minute inclusions can be seen at high magnifications but are too small to be resolved with objective lenses suitable for heating experiments. However, some useful data were obtained from inclusions in atypically coarse quartz from one telluride vein and from fluorite and quartz of telluride-bearing pyritic gold ore. The results are presented below after a brief summary of methods and assumptions.

Filling temperatures were measured with a conventional, homemade heating stage consisting of a small brass heating block with a central sample chamber and an outer nichrome heating element. The block is surrounded by a casing of asbestos with upper and lower Vycor windows for observation of the sample, and the entire unit is attached to the mechanical stage of a petrographic microscope. Temperatures are measured with a chromel-alumel thermocouple inserted through the block and its casing, and kept in direct contact with the sample during heating runs. Samples were prepared in the usual way by carefully grinding thin sections down to 1 mm or less and then polishing both surfaces to a high finish. The only problem encountered was the existence of thermal gradients within the central parts of the sample chamber, and for this reason much time was devoted to calibrating the instrument so that temperatures indicated by the centrally located thermocouple could be corrected for the true location of a specific inclusion within the cell. Commercially available "Tempil" powders of different melting points known to within 1 percent were used as standards for calibration. Once the gradient problem was solved, the only appreciable source of error (provided samples were kept thin) arose in judging the precise moment at which the inclusions became homogeneous. Repeated runs on the same inclusions moved to different locations within the cell gave corrected results within 5 degrees, and usually within 3 degrees for larger, plainly visible inclusions.

The errors involved in measuring filling temperatures are small compared to the pressure-salinity corrections that must be added to the results for true temperatures of crystallization. The writers are fairly confident that confining pressures during deposition of the ores were somewhere in the range 78 to 360 bars and, in order to refine the necessary corrections, some freezing experiments were performed to estimate salinities. These will not be discussed at length because they were largely unsuccessful. After various periods of time up to eight hours, in a dry ice-acetone bath only a few of the largest inclusions in each of the samples had frozen. On warming, these few inclusions thawed completely somewhere in the range -1° to -5°C, suggesting equivalent NaCl contents of 8 percent or less. Effervescence of acetone within the crude freezing stage obscured vision at the critical time of melting of ice in the inclusions and caused an erratic response from a thermistor inserted in the cell to record temperatures. Roedder (1962, 1963a) has successfully handled supercooling and equipment problems of this kind, but further work as part of the telluride project seemed impractical because so few samples contained workable fluid inclusions. For this reason, the experiments were abandoned, and in their place it was arbitrarily assumed that salinities of the inclusions vary from 0 to 20 weight percent NaCl (no visible NaCl appears in the inclusions at room temperature). Pressure-salinity corrections indicated by the experimental curves of Klevstov and Lemmlein (1959) were thus applied to the filling temperatures. In other words, the smallest possible correction of +8°C for pure water at 78 bars was added to the minimum filling temperatures while the largest possible correction of +30°C for a 20 percent NaCl solution at 360 bars was added to the maximum filling temperatures. The resulting temperature range was further expanded 5° at each extreme to include uncertainties of the measurements themselves.

The sources of error in such work are familiar to most economic geologists and will not be fully discussed here. Serious leakage problems were not encountered in this work and individual inclusions heated repeatedly gave the same filling temperatures within limits of observational error. Problems of prior natural leakage seem most unlikely, because filling temperatures of inclusions interpreted as primary within a given sample fall within a very narrow range that makes sense geologically; these check with other independent thermometers. We are least confident of the choices of true primary inclusions in the particular samples available for this work. The criteria applied in distinguishing primary and secondary inclusions are mentioned below.

One sample of telluride ore from the Gray Eagle mine contains quartz that is coarse enough for inclusion study. This quartz occurs as stubby crystals several millimeters long that contain scattered inclusions of pyrite and form comb-like linings along the borders of telluride veinlets. The biggest crystals contain many small but workable inclusions of the simple two-phase (water plus vapor) type. Pseudo-secondary or secondary inclusions are numerous, relatively small, and occur along planes that are oblique to obvious crystallographic planes. The primary (?) inclusions chosen for study do not obviously differ in contents from the secondaries, but are relatively large and remote from obvious secondary planes. Some are aligned or have faces aligned with external crystal faces, and many have straight walls aligned with extinction directions in the quartz. Some of these are negative hexagonal crystals, but most are of the irregular type shown in Plate 12f. No solids of any kind occur in the inclusions, and no visible CO_2 phase was produced in the freezing experiments. Upon heating, all inclusions filled by expansion of the liquid phase. Filling temperatures varied from sample to sample and from place to place in individual samples, but all were within the range 290° to 315°C. Correction of these results for pressure leads to estimated crystallization temperatures of 293° to 350°C. This is certainly a maximum temperature limit for tellurides in the Gray Eagle samples, and it can probably be regarded as a maximum for much finer-grained pyritic horn quartz that is more typical of the ores.

Fairly large inclusions were also found and tested in very coarse fluorite and quartz of pyritic gold ore from the Stanley mine near Jamestown, and the resulting data have an indirect but important bearing on the telluride mineralization. In the samples analyzed, coarse purple to violet fluorite is veined by coarse, glassy quartz containing large pyrite crystals of the pyritic gold stage of mineralization. In these samples intergrown bismuthinite, tetradymite, and free gold fill openings, and locally replaced the quartz and pyrite. This is the only recorded example of visible tellurides in pyritic gold ore. It is also significant that fragments of similar glassy quartz and purple fluorite occur in the Buena mine where they are cemented by horn quartz and ore minerals of the telluride stage of mineralization. These samples were analyzed on the assumption that they would also provide maximum temperatures for the later telluride mineralization.

Fluid inclusions in both the fluorite (Pl. 12g) and quartz (Pl. 12h) are also of the simple H_2O type (vapor plus fluid), contain no visible solids, and do not form a CO_2 phase upon cooling. The minerals lack growth zoning and primary inclusions could not be selected on the basis of their relations to external or internal growth features although inclusions chosen in the quartz are aligned with extinction directions. No helpful compositional differences among different inclusions were noted and so the choice of primary inclusions was an uncertain one. Filling temperatures were measured for 20 of the largest inclusions in the fluorite that were not obviously aligned in secondary planes, and these all fell in the range 275° to 320°C. Twelve primary (?) inclusions in the younger quartz filled in the range 237° to 255°C. The apparent depositional temperatures corrected for pressure are 278° to 355°C for the fluorite and 240° to 285°C for the younger quartz. Heating runs were carried beyond the filling temperature and this caused plainly audible decrepitation in the fluorite at about 340°C, and at about 360° the thin sections either moved or split.

Blanchard (1954) has studied the thermoluminescence of fluorites from the Jamestown district, including samples of the purple fluorite from the Stanley mine. He concludes that thermoluminescent intensities in a given generation of fluorite are related to proximity to the Porphyry Mountain stock. Thermoluminescent curves (heating rate 10°C/minute) of the Stanley fluorite show a first glow at about 110°C, maximum intensity in the range

375° to 480°C, and a final glow at 515° to 560°C. Very probably, the purple radiation colors which are related to small amounts of uraninite in the samples (Goddard, 1946) could be completely destroyed by prolonged heating at temperatures at or below 110°C, and therefore well below the depositional temperatures indicated by the fluid inclusions. We believe that the relationship between glow intensities and proximities to the Porphyry Mountain stock may reflect an original control of radioactive element contents in the fluorite by temperature, but the glow curves cannot be used to determine the depositional temperatures directly.

Summary

Temperatures indicated by these different thermometers are summarized diagrammatically in Figure 27 where the inferred depositional temperatures for various groups of minerals precipitated in the telluride stage of mineralization are also shown. Considering the number and variety of thermometers applied here, there is a remarkable degree of agreement among them. The earliest stages of mineralization, as represented by unusually coarse pyritic quartz of the Gray Eagle mine and perhaps by the arsenopyrite-bearing quartz of the Archer mine, started at temperatures locally as high as 350°C. However, there are several indications that the bulk of the early pyritic quartz formed below 300°C (K-feldspar+kaolinite in the wall rocks), and very probably below 250°C (below depositional temperatures of coarse quartz in the pyritic gold ores). At the lower end of the scale, there are many indications (maxima for goethite, bravoite, sylvanite-hessite, sylvanite-petzite) that precipitation of the tellurides started or at least persisted below 170°C. The inversion of intermediate to low hessite at 145°C in the binary system Ag-Te is indicated in Figure 27, but the uncertainties of this thermometer have already been stressed. The lowest depositional temperatures, probably reached during formation of late native gold, are unknown but were probably on the order of 100°C. This would be very close to temperatures of 70° to 100°C that would be expected at the estimated depths of mineralization in an active orogenic region of high thermal gradient (ca. 50°C per kilometer, depths 2600 to 4600 feet, surface at mean 32°C). Some features of the ores such as the breakdown of the metastable "x" phase to petzite-hessite intergrowths probably occurred long after the time of active mineralization as the geothermal gradient returned to normal. Very few of the thermometers enable us to break down events within the entire vein sequence, and so the temperature ranges indicated for native gold and the sulfides are inferred from the known ages relations of these minerals with respect to pyritic quartz and the tellurides. In summary, there is ample evidence that the telluride mineralization began at temperatures locally as high as 350°C but generally below 250°C and continued as the temperatures fell to about 100°C. At any one time, the environment probably was laterally isothermal, but in any one place, the temperatures declined through time as the mineralization progressed.

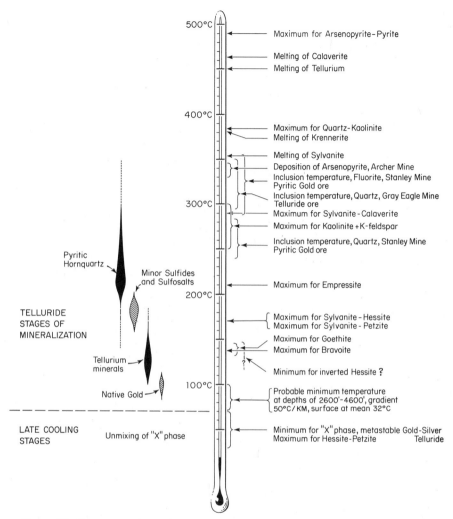

Figure 27. Summary of geothermometers and estimated depositional temperatures of the telluride stage of mineralization.

SOURCES OF THE METALS AND OTHER VEIN COMPONENTS

Lovering (1941) presented a very strong case for derivation of the telluride ore fluids from a biotite latite intrusive source thought to underlie the telluride belt. With such an origin for the fluids in mind, it is of interest to speculate on sources of the metals and other vein components.

Very clearly, some important vein constituents were picked up by the ascending fluids in their reactions with wall rocks adjacent to the channelways. Available chemical analyses of the fresh and altered wall rocks (Lovering and Tweto, 1953; Bonorino, 1959) show that the fluids donated H^+ and K^+ to the walls, and in the process acquired Si, Fe, Ca, Mg, and Na. The importance of this exchange is that it might account entirely for the Fe and Si of pyritic horn quartz which actually makes up most of the vein matter; only sulfur would of necessity be imported by the fluids to form pyrite. In addition, this process could have supplied the small amounts of Ca and Mg in the calcite and ankerite-dolomite that are locally abundant as vein minerals. Spectrographic analyses (Bray, 1942a) have shown that the Boulder Creek and Silver Plume granites contain moderate to large traces of barium and vanadium, but these elements have not been studied in detail in the alteration envelopes. Presumably, the wall rock alteration may have released quantities sufficient to account for roscoelite and minor barite in the telluride veins.

Bray's spectrographic work (1942a and b) also showed that Ag, Zn, Pb, and Ni occur in minor quantities in the Precambrian granites and some of the Early Tertiary intrusives that predated the telluride mineralization, but there is no evidence that there was any transfer of these metals to the veins during the mineralization, except for very small amounts that might have been released from the narrow alteration envelopes themselves.

A final source to be considered is the older lead-silver and pyritic gold deposits, some of which are exposed and others which may be concealed at depth along the channels followed by the telluride fluids in their ascent. Hydrothermal leaching of such deposits might release significant amounts of Au, Ag, Pb, Zn, Sb, S, and other components for reprecipitation in the telluride veins. Although this possibility is intriguing, there is little evidence to support it. The exposed lead-silver and pyritic gold ores show no cavities, boxworks, or other solution features suggestive of hypogene leaching and in the Jamestown district it appears that the telluride solutions were actually diverted away from ground previously sealed or grouted during earlier stages of mineralization. The widespread occurrence of accessory melonite in the telluride veins suggests the possibility that nickel may have been leached at depth from older Precambrian deposits such as that at the Copper King mine near Gold Hill (Goddard and Lovering, 1942). However, there is no evidence of hydrothermal leaching of this deposit and the abundance and distribution of nickel in the telluride veins bear no obvious relationship to this older nickel occurrence (see Fig. 13). In summary, the writers can neither prove nor disprove reworking of older deposits at depth, but see no signs of this in deposits exposed at the surface.

Except for important contributions of Si, Fe, Ca, Mg, and possibly V and Ba by the altered wall rocks, the other components of the telluride veins including the base and precious metals, sulfur, and tellurium, are thought to have come from the biotite latite source at depth. The latite itself and its related telluride ores were part of a grander sequence of Early Tertiary intrusives and associated ore deposits derived by gradual differentiation of more extensive parent melts underlying this part of the Front Range mineral belt (Lovering, 1941; Lovering and Goddard, 1938, 1950).

UNSOLVED PROBLEMS

Some aspects of the telluride mineralization are important, but poorly understood. For example, the general question of the mode of transport of the metals is fundamental and has been debated for many years without being resolved. No fresh insights to this classical problem are offered by the present study but, if anything, the precipitation of a complex group of tellurides from fluids that also produced a suite of common sulfides seems to place new demands on hydrothermal solvents that are already overworked. The presence of tellurium in these fluids raises interesting new questions as to whether bitelluride or perhaps polytelluride ions might play roles analogous to those envisaged by some for bisulfide or polysulfide ions. Should one be thinking in terms of a bisulfide or polysulfide solution carrying tellurium itself as a soluble sulfur complex? There has been virtually no experimentation in the system $Te-H_2O$ at elevated temperatures, but the $Eh-pH$ relations at room temperature (see Fig. 21; Deltombe and others, 1956) would seem to rule out any important role for the bitelluride ion which is stable only under extremely reducing conditions. As a matter of fact, most of the ions that might be considered as possible carriers, such as $Te_2^=$, Te^-, $HTeO_3^-$, and TeO_3^-, are unstable with respect to solid tellurium or tellurium oxide over the range of acidities and oxidation potentials to be expected in hydrothermal environments. In this respect, tellurium may be very different from sulfur. The existence of unknown polytellurides at high temperatures cannot be entirely ruled out but it is reasonable to expect that these would be greatly subordinate to other species like $Te_2^=$, $HTeO_3^-$, or TeO_3^-, which themselves would be of very low activity.

The present writers favor the possibility that tellurium is transported along with the metals in solution as a soluble chloride complex. Although its precise formula is unknown, at least one tellurium chloride complex ($TeCl_6^=$) is known to be stable at room temperature and probably persists under hydrothermal conditions. The telluride fluids are very likely chloride solutions carrying a great variety of ions like $TeCl_6^=$, $AuCl_4^-$, $AgCl_2^-$, and $AgCl_4^=$, along with chloride complexes of the more common base metals. This view calls for the least modification or complication of any classical theory in order to explain the precipitation of both tellurides and sulfides from the same ore fluids. The telluride solutions are in this case unique only because of their unusual abundance of tellurium which otherwise is carried along in complex chloride form like the other cations. Such a view is consistent with other recent studies of geothermal brines and fluid inclusions that indicate a high chloride, low sulfur content in fluids known or reasonably presumed to be ore-forming fluids.

Another difficult problem is that of estimating the time required for the telluride mineralization. Tweto (1947) gave considerable thought to this problem in his studies of the tungsten ores of Boulder County and evidently felt, at the time, that the long sequence of horn quartz types, the record of successive vein movements, and the wide alteration zones (up to 50 feet in places) found along the tungsten veins called for a prolonged period of mineralization. Although the alteration envelopes are much narrower along the telluride veins, his sound arguments seem applicable to the telluride mineralization as well. However, these time-dependent effects might be variously interpreted and depend themselves on other imponderables. In view of sluggish reaction rates of silicates in hydrothermal experiments performed at relatively low temperatures, the writers feel that extensive wall rock alteration is evidence for a long-lived process taking perhaps hundreds or even thousands of years. On the other hand, actual rates of alteration are poorly known and might be surprisingly rapid if all involved factors such as the concentrations of various cations in the ore fluids were favorable. With respect to the structural movements and horn quartz sequence within the veins, it would not be difficult to visualize the necessary structural events occurring with almost catastrophic rapidity in an active orogenic belt, but the critical question is whether precipitation could be fast enough to account for continued mineralization of all the new vein openings. Any judgment here calls for assumptions regarding the concentrations and flow velocities of the ascending ore fluids. As pointed out by Roedder (1962, 1963a and b), the character of fluid inclusions trapped in vein minerals indicates slowly moving ore fluids. If the fluids were moving rapidly through newly brecciated and gougey channels, one should expect to see some sort of fine-grained, mechanically transported debris trapped along with the solutions. However, no solids of any kind occur in fluid inclusions of vein quartz from the telluride deposits. As a matter of fact, the problems of extreme supercooling encountered in attempting to freeze inclusions in quartz from the Gray Eagle mine are very good evidence of a complete lack of solid particles that would serve to nucleate ice (Roedder, 1963a). Still another feature of the Boulder deposits that seems to call for slowly moving ore fluids is the character of the telluride crystals themselves. Many slender, fragile crystals of tellurium, sylvanite, and calaverite with mirror-like, perfect faces are exposed in vuggy openings along the veins. It is difficult to see how such crystals could lie in the path of rapidly moving, turbid ore fluids without undergoing some degree of abrasion or even breakage. For these various reasons the authors feel that the telluride ores must have formed from slowly moving ore fluids, and that the process was a very slow one requiring probably hundreds of years or more.

Lovering (1941) also appraised the time element in connection with the tungsten deposits of Boulder County which he and later Tweto (Tweto, 1947; Lovering and Tweto, 1953) believed were derived along with the telluride ores from the biotite latite source at depth. The present authors do not share any conviction that the tungsten ores were also derived from this source, but if this possibility is to be entertained, then it is difficult to divorce any considerations of the tungsten mineralization from the topic at hand. Lovering called attention to the high water content (4.06 H_2O+ and

4.60 H_2O-) of the biotite latite rocks and the persistence of thin seams of breccia-charged latite as evidence of a tremendous vapor pressure in the magmatic reservoir and a sudden, explosive release of the mineralizing volatiles. It is here suggested that the high water content of these rocks merely proves the minimum water content of the original magma, and one could argue (equivocally) that the fact that the water is still in the rocks is evidence that there was not even a normal, much less an explosive, release of volatiles from the magma. The physical character of the intrusion breccia is more convincing of a violent origin, but the breccia is only one, and perhaps a minor, phase of the biotite latite intrusion, and it would be unsafe to judge the character of the main biotite latite source at depth on the basis of a few small bodies of shallow breccia close to the present surface.

In their study of the tungsten district, Lovering and Tweto (1953) offered an interesting theory of mineralization in which the timing and duration of the tungsten mineralization are intimately involved with their deductions as to the changing character of the ore fluids. The theory was a sound and intriguing one in keeping with the stability relations of the alteration and vein minerals as understood at the time, but the writers feel this theory should be re-examined in the light of recent experimental studies not available to Lovering and Tweto. Such re-examination would require a thorough investigation of alteration types associated with each of the different stages of mineralization in Boulder County.

Classification

MINERALOGICAL BASES FOR CLASSIFICATION
OF TELLURIDE DEPOSITS

If the mineralogy of telluride ores were governed chiefly by environmental factors rather than compositional properties of the ore fluids, this would obviously offer a basis for genetic classification of these unusual deposits. Several researchers have explored this possibility by attempting to relate the occurrences of certain tellurium minerals or groups of tellurides to the geological settings in which they occur. Helke (1934) suggested that tellurides other than those of gold and silver are characteristic of high temperature settings and are sparse or lacking in epithermal ores. He also refuted the idea, widespread at the time, that the gold tellurides are concentrated in epithermal deposits. Ishibashi (1960) relates the combining tendencies of tellurium with gold and bismuth to depths of mineralization proposing that Te combines with Bi rather than Au in plutonic hydrothermal settings, but with Au rather than Bi in volcanic hydrothermal deposits. Markham (1960) distinguishes three basic types of telluride mineralization based in part on differences of tellurium mineralogy. These are (1) deposits associated with Tertiary volcanics formed at relatively low temperatures and pressures, (2) Precambrian deposits formed at relatively high temperatures and pressures, and (3) minor bismuth tellurides in contact metamorphic deposits. The last category, represented by deposits at Hedley, British Columbia, the Oya mine (Japan), and Glassford Creek (Queensland) are interesting but quantitatively unimportant so far as production of tellurides is concerned. Placing great emphasis on Kalgoorlie and Vatukoula as representatives of the two important environments, Markham concludes that the Tertiary volcanic settings are characterized by an abundance of native tellurium, rarity of free gold, predominance of krennerite over calaverite, and a paucity of tellurides other than those of gold and silver. Conversely, the deeper-seated Precambrian deposits are reportedly typified by abundant native gold, rarity of uncombined tellurium, a predominance of calaverite over krennerite, and the common occurrence of tellurides of lead, mercury, copper, and bismuth. On the basis of his experimental work,

Markham thought that krennerite was the low-temperature, low-pressure polymorph of $AuTe_2$, and thus explained occurrences of krennerite in shallow, low-temperature settings where he also inferred that krennerite predominates over calaverite.

Callow and Worley (1965) had difficulties in attempting to fit the telluride deposits of the Acupan mine (Philippine Islands) into Markham's classification. Geologically, these deposits belong in the Tertiary volcanic category, but free gold is common, tellurium absent, and calaverite predominates over krennerite. Cabri (1965b) has disproved the polymorphic relationship between calaverite and krennerite, and stressed the compositional controls of occurrence of these two distinct minerals. He further proposed that the differences between tellurium-rich assemblages of the Vatukoula type and native gold-rich assemblages of the Kalgoorlie type are also functions of bulk composition of the ore-forming fluids and not necessarily of temperature, pressure, or other environmental factors (Cabri, 1965a and b).

Many of the generalizations that have been made with regard to selective environmental occurrence of the tellurides do not hold up when a sufficient number of world districts are considered. This point is illustrated by Table 14 where the relative abundances of native gold, native tellurium, and tellurides in only ten well-known telluride districts are listed. The abundances listed for these minerals are relative to one another and overlook the perhaps significant abundances of associated sulfides which vary greatly among these localities. The data listed for the British Columbia "district" refer collectively to many mines that have yielded tellurides in a belt extending along the eastern margin of the Coast Range batholith from 140 to 240 miles north of Vancouver. The abundances listed for this broad belt are based mostly upon the numerous X-ray identifications published by Thompson (1949) and recent personal communications with Thompson (1966). The other districts require no special comment because they are more obvious and perhaps more familiar geographically. It should be noted that two possible classifications are given for Precambrian gold deposits of the Kirkland Lake district, Ontario, in Table 14. These deposits, long considered representatives of the mesothermal class (Todd, 1928; Lindgren, 1933; Hawley, 1948), have recently been cited as examples of metamorphosed deposits initially formed as integral parts of a volcanic complex (Goodwin, 1961, 1962, 1965). Although these different classifications imply a very different sequence of events in the history of the ores, both would call for epigenetic emplacement of the ores to their *present* sites and for the formation of the *present* mineral assemblages under high P-T conditions in a deep seated environment.

Compilations of the type presented in Table 14 call upon sources that differ in vintage and quality and are never quite satisfactory, but they serve as a basis for evaluating categorical statements that relate the tellurides with specific environments. If one accepts the published identifications

TABLE 14: RELATIVE ABUNDANCES OF TELLURIDES AND NATIVE METALS IN SELECTED MINING DISTRICTS OF THE WORLD

DISTRICT	CLASSIFICATION	AGE	Free Gold	Free Tellurium	Altaite	Calaverite	Coloradoite	Empressite	Hessite	Krennerite	Melonite	Nagyagite	Petzite	Stuetzite	Sylvanite	Tetradymite	Other Bi-tellurides	Sources of Data
Cripple Creek, Colorado	Epithermal	Tertiary	○	○		●			○	●			○		◐			Loughlin & Koschmann (1935) Lindgren & Ransome (1906) Tunell (1954)
Nagyag, Rumania Karolinen Terrain Longin Terrain	Epithermal	Tertiary	○ ○ ○		○				○	○ ●		● ○	○ ○ ○	○	○ ● ●			Helke (1934)
Vatukoula, Fiji Islands	Epithermal	Tertiary		●		●	○		○				○		○			Stillwell (1940) Markham (1960)
Goldfield, Nevada	Epithermal	Tertiary	●			●			○	●			○		○			Tolman & Ambrose (1934) Ransome (1909, 1910) Searls (1948)
Acupan mine Phillipine Is.	Epithermal	Tertiary	●	○	●	◐	○	○	●			○	●	○	○	○		Callow & Worley (1965)
Boulder County Colorado	Epithermal	Tertiary	●	○	●	◐	●	○	◐	◐	◐	○	●	○	●	○		Present Paper
British Columbia	Mesothermal	Mesozoic	●		●	○	◐		○		○		◐		○	●	●	Thompson (1949; personal communication, 1966) Warren (1947)
La Plata, Colorado	Mesothermal	Tertiary	●	○	●	○	●			●			○		○			Galbraith (1941) Eckel and others (1949)
Kirkland Lake Ontario	Mesothermal (Metamorphosed Volcanic?)	Precambrian	●		●	●	○		◐		○		●		○	○	○	Todd (1928) Goodwin (1961, 1962, 1965) Hawley (1948) Thompson (1949)
Kalgoorlie, Australia	Mesothermal-Hypothermal?	Precambrian	●	○	○	●	●		◐	◐	○	○	◐	○	◐	◐		Stillwell (1931) Markham (1960)

● Primary Abundance "Abundant," "Chief," "Major" ...
◐ Secondary Abundance "Common," "Subordinate" ...
○ Tertiary Abundance "Minor," "Rare," "Reported," etc.

and classifications given in Table 14, it is first of all apparent that the proportions of calaverite and krennerite in a given district are a very poor index of environment. Within the epithermal category alone, these minerals may be equally abundant (Cripple Creek, Tunell, 1954) or either calaverite (Acupan and Boulder) or krennerite (Vatukoula) may predominate. Tellurium is uncommon in deeper-seated deposits but either free gold or native tellurium may predominate in shallow ores and in rare cases (Boulder County belt) both are abundant in the same region. Generalizations regarding occurrences of tellurides of lead, copper, and mercury are not firmly supported by the record. The copper tellurides are, with rare exceptions (e.g., Good Hope mine, Vulcan, Colorado), supergene and have no place in the classification of primary deposits. The tellurides of lead and mercury occur abundantly in very different settings, and other tellurides like melonite $NiTe_2$ have been recognized in so few districts that their modes of occurrence are not well established. There does seem to be a tendency for the bismuth tellurides to occur preferentially in deposits thought to have formed at greater depths. While tetradymite has been recorded in shallow settings (e.g., Boulder County belt), this mineral along with other bismuth tellurides like wehrlite, tellurbismuth, and so forth (see Thompson, 1949) seem to be most common in mesothermal and even more intense settings. In this connection, it is interesting that tetradymite is one of the few tellurides recorded in the Sudbury deposits (Hawley, 1962, pp. 102-103) and, as noted by Markham (1960), the bismuth tellurides are among the few that have been found in contact metamorphic deposits. A more thorough review of other deposits might bear out this apparent environmental control of the bismuth tellurides.

At present, there does not appear to be any consistent scheme of classification of telluride deposits inherent in the tellurium mineralogy *per se*. As recognized by Markham (1960), most but not all of the known deposits are either Tertiary volcanic (or sub-volcanic) or belong to a second category of relatively deep Precambrian ores. However, a given deposit cannot be confidently placed in either of these categories on the basis of the tellurides and native elements that are present.

THE BOULDER COUNTY DEPOSITS

The telluride ores of Boulder County are an example of complex Tertiary mineralization at low temperatures and fairly shallow depths in Precambrian terrain. Erosion has removed any volcanics erupted at the time of mineralization, but has, in the process, exposed dikes and intrusion breccias of biotite latite that are both temporally and spatially related to the ores. A large body or perhaps several large intrusives of similar composition are thought to lie beneath the telluride belt at unknown depths. The known ores were probably formed at depths of about 2600 to 4600 feet below the original surface. Local depositional temperatures as high as 350°C are indicated, but most of the ore is thought to have formed in the range 100° to 250°. Although some of the early pyritic vein quartz formed at temperatures somewhat above those specified for the epithermal environment by Lindgren (1933), the deposits are otherwise an excellent example of this class of deposit.

Compared mineralogically with other telluride districts, the Boulder ores are unusual in two respects. First, they contain an exceptional variety of vein minerals which are not only numerous, but make up significant parts of the telluride assemblages. Second, this is the only major district in which both the native tellurium-rich and native gold-rich types of ore are well represented. The ores of separate productive centers of this belt are as different from one another as are those of widely separated mining districts and, in a sense, one might liken the tellurium-rich John Jay center to Vatukoula, the petzite-altaite-free gold ores of the Smuggler center to the Acupan mine, and draw other similar parallels to other districts of the world. As a whole, however, the Boulder belt is mineralogically dissimilar to any other single camp. In view of the evidence that most of the ores throughout this belt have formed at approximately the same depths and temperatures, the authors suspect that the unusual features noted above are due to compositional rather than environmental causes. In previous pages the writers presented the idea of compositionally distinctive fluids mineral-izing the separate structural centers of the telluride belt which would so explain the presence of varied and extreme types of ore within such a geographically restricted area.

Conclusions

The Boulder County deposits have provided an unusual opportunity to study telluride ores produced in a single stage of hydrothermal deposition and subsequent cooling without any serious complications of post-depositional metamorphism, deformation, or weathering. Compared to other world telluride localities, this small area in Colorado has produced ores of remarkable mineralogic variety; sixty-seven vein minerals have been identified including 15 tellurides, 15 sulfides, and 5 sulfosalts, as well as native gold and native tellurium. In addition to documenting this variety, the present study has shown that this variety is caused chiefly by differences in the bulk composition of the fluids that mineralized the separate productive centers of the telluride belt rather than by drastic differences of depositional temperature or confining pressure. The fluids are thought to have inherited these differences from a biotite latite source at depth and these were preserved to varying degrees as the fluids ascended the major breccia reef structures. The reefs have therefore played a major role not only in determining the spatial localization of ore, but also in affecting its ultimate mineralogic composition.

The original hypogene textures and associations of the telluride ores have been extensively modified by cooling, and the original hypogene sequence can be reconstructed only if detailed microscopic observations are coordinated with application of pertinent experimental phase relations. A high degree of local equilibrium was evidently maintained during the initial deposition of the tellurides, but subsequent re-equilibration produced by cooling seems to have varied among ores of different bulk composition. Certain puzzling intergrowths of sylvanite, hessite, and petzite, commonly found in Boulder County and in ores of many other telluride districts, evidently form upon cooling and decomposition of unstable minerals once present in the veins. Some of these changes may take place long after the period of active mineralization as the mineralized terrane slowly cools.

Boulder County also provides an excellent testing ground for geothermometers applicable to telluride ores. Based on physiographic evidence, the ores are thought to have formed at depths of from 2600 to 4600 feet and at confining pressures in the range of 78 to 360 bars. These figures closely

189

agree with those given by the arsenopyrite barometer, and can therefore be used with some confidence in correcting any temperature estimates. The many geothermometers applied give surprisingly consistent results and indicate that, at any point in the veins, the depositional temperatures declined through time. In the earliest stages of telluride mineralization, the temperatures were locally as high as 350°C, but the bulk of the vein material probably formed in the range of 100° to 250°C.

The behavior of tellurium, iron, gold, and silver during oxidation of the Boulder ores is interpreted in terms of Eh-pH and Eh-a Cl− equilibrium diagrams computed for tellurium-bearing systems. These theoretical considerations support the probable occurrence of supergene native tellurium as first inferred from field evidence. They also suggest that, in some telluride deposits, hessite could have played a role analogous to that of supergene chalcocite or argentite in enriched sulfide deposits.

A brief review of some of the world's major telluride deposits shows that there is no reliable scheme of genetic classification that can be based on the tellurium mineralogy *per se*. With the possible exception of the bismuth tellurides, almost all of the more common tellurides can be found in a variety of geologic settings and are, therefore, poor indicators of their formational environment. Based on the other considerations, the Boulder County telluride deposits are best classified in the epithermal class of the traditional intensity scale. They provide an excellent example of complex Early Tertiary mineralization in Precambrian terrane.

References Cited

Argall, G. O., Jr., (1943), Scheelite occurrences in Colorado: Mines Mag., v. 33, p. 313-314.

Axelrod, J. M., (1946), A field test for vanadium: U.S. Geol. Survey Bull. 950, p. 19-23.

Baker, G., (1958), Tellurides and selenides in the Phantom Lode, Great Boulder mine, Kalgoorlie: Australasian Inst. Mining Metall., Stillwell Volume, p. 15-40.

Becking, L. G. M. B., Kaplan, I. R., and Moore, D., (1960), Limits of the natural environment in terms of pH and oxidation-reduction potentials: Jour. Geology, v. 68, p. 243-284.

Berek, M., (1937), Optische Messmethoden in polarisierten Auflicht: Fortschritte du Mineralogie, Petrographie, und Kristallographie, v. 22, p. 1-104.

Berry, L. G., and Thompson, R. M., (1962), X-ray powder data for ore minerals: Geol. Soc. America Memoir 85, 281 p.

Bjerrum, J., Schwarzenbach, G., and Sillen, L., (1958), Stability constants; Part II. Inorganic ligands: Chem. Soc. (London), Spec. Paper 7, 131 p.

Blanchard, F. N., (1954), Thermoluminescent properties in relation to geologic occurrence: Unpublished Master's Thesis, Dept. Geology, Univ. Michigan, Ann Arbor, 51 p.

Bonorino, F. G., (1959), Hydrothermal alteration in the Front Range mineral belt: Geol. Soc. America Bull., v. 70, p. 53-90.

Borchert, H., (1935), Neue Beobachtungen an Tellurirzen: Neues Jahrb. fur Mineralogie, Beil. Bd. 69A, p. 460-477.

Bowie, S. H. V., (1962), Reflection characteristics of ore minerals (Discussion): Econ. Geology, v. 57, p. 983-985.

——, (1965), Minutes of the I.M.A. Commission on Ore Microscopy held in New Delhi on December 14, 1964: Econ. Geology, v. 60, p. 1326-1328.

Bowie, S. H. V., and Taylor, K., (1958), A system of ore mineral identification: Mining Mag., v. 99, p. 265-277, 337-345.

Bray, J. M., (1942a), Spectroscopic distribution of minor elements in igneous rocks from Jamestown, Colorado: Geol. Soc. America Bull., v. 53, p. 765-814.

——, (1942b), Distribution of minor chemical elements in Tertiary dike rocks of the Front Range, Colorado: Am. Mineralogist, v. 27, p. 425-440.

Cabri, L. J., (1965a), The occurrence of telluride minerals at the Acupan gold mine, Mountain Province, Philippines (Discussion): Econ. Geology, v. 60, p. 1080-1082.

——, (1965b), Phase relations in the system Au-Ag-Te and their mineralogical significance: Econ. Geology, v. 60, p. 1569-1606.

——, (1965c), Empressite and stuetzite redefined (Discussion) : Am. Mineralogist, v. 50, p. 795-801.

——, (1967), Note on the occurrence of calaverite and petzite in the Phantom Lode, Great Boulder mine, Kalgoorlie; Australasian Inst. Mining Metall., Proceedings, no. 222, p. 95.

Callow, K. J., and Worley, B. W., Jr., (1965), The occurrence of telluride minerals at the Acupan gold mine, Mountain Province, Philippines: Econ. Geology, v. 60, p. 251-268.

Cameron, E. N., (1961), Ore microscopy: New York, John Wiley, 293 p.

——, (1963), Optical symmetry from reflectivity measurements: Am. Mineralogist, v. 48, p. 1070-1079.

Cameron, E. N., Davis, J. H., Guilbert, J. M., Larson, L. T., Mancuso, J. J., and Sorem, R. K., (1961), Rotation properties of certain anisotropic ore minerals: Econ. Geology, v. 56, p. 569-583.

Cameron, E. N., and Threadgold, I. M., (1961), Vulcanite, a new copper telluride from Colorado: Am. Mineralogist, v. 46, p. 258-268.

Carpenter, R. H., and Cameron, E. N., (1963), Additional measurements of rotation properties of ore minerals: Econ. Geology, v. 58, p. 1309-1312.

Chace, F. M., (1956), Abbrevations in field and mine geological mapping: Econ. Geology, v. 51, p. 712-723.

Christie, O. H. J., (1962), Unit cell and space group of mercury tellurate, Hg_3TeO_6: Acta Cryst., v. 15, p. 94-95.

Clark, F. W., (1877), An analysis of sylvanite from Colorado: Amer. Jour. Sci., 3d ser., v. 14, p. 286.

Clark, L. A., (1960a), The Fe-As-S system. Part I. Phase relations and applications: Econ. Geology, v. 55, p. 1345-1381.

——, (1960b), The Fe-As-S system: variations of arsenopyrite composition as function of T and P: Carnegie Inst. Washington Yearbook 59, p. 127-130.

Cloke, P. L., and Kelly, W. C., (1964), Solubility of gold under inorganic supergene conditions: Econ. Geology, v. 59, p. 259-270.

Dana, E. S., (1892), The system of mineralogy of James Dwight Dana, Descriptive mineralogy: New York, John Wiley, 6th ed., 1134 p.

Deltombe, E., de Zoubov, N., and Pourbaix, M., (1956), Comportement électrochimique du tellure. Diagram d'équilibre tension-pH du system Te-H_2O à 25°C: Centre Belge d'Étude de la Corrosion, Rapport Technique 33, p. 22.

Eckel, E. B., (1949), Geology and ore deposits of the La Plata district, Colorado: U.S. Geol. Survey Prof. Paper 219, 179 p. (*with sections by* Williams, J. S., Galbraith, F. W., and others).

——, (1961), Minerals of Colorado, a 100-year record: U.S. Geol. Survey Bull. 1114, 399 p.

Edwards, A. B., (1954), Textures of the ore minerals: Melbourne, Australasian Inst. Mining Metall., 155 p.

Emmons, W. H., (1917), The enrichment of ore deposits: U.S. Geol. Survey Bull. 625, 530 p.

Endlich, F. M., (1874), Tellurium ores of Colorado: Eng. and Mining Jour., v. 18, p. 133.

———, (1878), Mineralogical report: Hayden Survey: 10th Ann. Rept., 1876, pp. 133-159.

Frondel, C., and Pough, F. H., (1944), Two new tellurites of iron; mackayite and blakeite. With new data on emmonsite and "durdenite": Am. Mineralogist, v. 29, p. 211-225.

Frueh, A. J., Jr., (1959a), The crystallography of petzite, Ag_3AuTe_2: Am. Mineralogist, v. 44, p. 693-701.

———, (1959b), The structure of hessite, Ag_2Te-III: Zeitschr. Kristallographie, v. 112, p. 44-52.

Furbish, W. J., (1963), Geologic implications of jarosite, pseudomorphic after pyrite: Am. Mineralogist, v. 48, p. 703-706.

Galbraith, F. W., 3rd, (1940), Identification of the common tellurides: Am. Mineralogist, v. 25, p. 368-371.

———, (1941), Ore minerals of the La Plata mountains, compared with other telluride districts: Econ. Geology, v. 36, p. 324-334.

Galopin, R., (1947), Differentiation chemique des minéraux métalliques par la méthode des empreintes: Schweizer. Mineralog. u. Petrog. Mitt., v. 27, p. 190-235.

Garrels, R. M., (1960), Mineral equilibria: New York, Harper and Row, 254 p.

Garrels, R. M., and Christ, C. L., (1965), Solutions, minerals and equilibria: New York, Harper and Row, 450 p.

Genth, F. A., (1874), On American tellurium and bismuth minerals: Am. Phil. Soc. Proc., v. 14, p. 223-231.

———, (1877), On some tellurium and vanadium minerals: Am. Philos. Soc. Proc., v. 17, p. 113-123.

Goddard, E. N., (1935), The influence of Tertiary intrusive structural features on mineral deposits at Jamestown, Colorado: Econ. Geology, v. 30, p. 370-386.

———, (1940), Preliminary report on the Gold Hill mining district, Boulder County, Colorado: Colorado Sci. Soc. Proc., v. 14, p. 103-139.

———, (1946), Fluorspar deposits of the Jamestown district, Boulder County, Colorado: Colorado Sci. Soc. Proc., v. 15, p. 1-47.

Goddard, E. N., and Lovering, T. S., (1942), Nickel deposit near Gold Hill, Boulder County, Colorado: U.S. Geol. Survey Bull. 931-O, p. 349-362.

Goodwin, A. M., (1961), Some aspects of Archaen structure and mineralization: Econ. Geology, v. 56, p. 897-916.

———, (1962), Volcanic complexes and mineralization in northeastern Ontario: Canadian Mining Jour., v. 83, no. 4, p. 62-65.

———, (1965), Mineralized volcanic complexes in the Porcupine-Kirkland Lake-Noranda region, Canada: Econ. Geology, v. 60, p. 955-971.

Guiteras, J. R., (1937), Operations and costs at the St. Joe Mining and Milling Co., Boulder County, Colorado: U.S. Bur. Mines Inf. Circ. 6976, p. 31.

Gutzeit, G., (1942), Determination and localization of metallic minerals by the contact print method: Am. Inst. Mining Metall. Engineers Tech. Publ. 1457, 13 p.

Hallimond, A. F., (1953), The polarizing microscope: York (England), Cooke, Troughton, and Simms, 2d ed., 204 p.

Hase, D. R., (1952), The application of polarization figures and rotation properties to the identification of certain telluride minerals: Econ. Geology, v. 47, p. 807-814.

Hawley, J. E., (1948), Mineralogy of the Kirkland Lake ores: Ontario Dept. Mines, v. 57, p. 104-124.

———, (1962), The Sudbury ores: their mineralogy and origin: Canadian Mineralogist, v. 7, p. 1-207.

Headden, W. P., (1903), Mineralogical notes: Colorado Sci. Soc. Proc., v. 7, p. 141-150.

Heinrich, E. W., and Levinson, A. A., (1955), Studies in the mica group; X-ray data on roscoelite and barium muscovite: Am. Jour. Sci., v. 253, p. 39-131.

Heinrich, E. W., Levinson, A. A., Levandowski, D. W., and Hewitt, C. H., (1953), Studies in the natural history of micas: Univ. Michigan, Engineering Research Inst. Final Rept. Proj. M978, 241 p.

Helke, A., (1934), Die Goldtellurerzlägerstatten von Sacaramb (Nagyag) in Rumänien: Neues Jahrb. fur Mineralogie, Beil. Bd. 68A, p. 19-85.

Hemley, J. J., and Jones, W. R., (1964), Chemical aspects of hydrothermal alteration with emphasis on hydrogen metasomatism: Econ. Geology, v. 59, p. 538-569.

Hess, F. L., and Schaller, W. T., (1914), Colorado ferberite and the wolframite series: U.S. Geol. Survey Bull. 583, 75 p.

Hiemstra, S. Q., (1956), An easy method to obtain X-ray diffraction patterns of small amounts of material: Am. Mineralogist, v. 41, p. 519-521.

Hillebrand, W. F., (1885), Mineralogical notes, Colorado Sci. Soc. Proc., v. 1, p. 121-123.

Hodge, E. T., (1915), The composition of waters in mines of sulfide ores: Econ. Geology, v. 10, p. 123-139.

Holser, W. T., (1953), Limonite is goethite: Acta Cryst., v. 6, p. 565.

Honea, R. M., (1964), Empressite and stuetzite redefined: Am. Mineralogist, v. 49, p. 325-338.

Ingersoll, L. R., Zobell, O. J., and Ingersoll, A. C., (1954), Heat conduction; with engineering, geological, and other applications: Madison (Wisconsin), Univ. Wisconsin Press, 325 p.

Ishibashi, M., (1960), Au-Ag tellurides from that Daté mine, Hokkaido, Japan: Faculty Engineering, Hokkaido Univ., Memoir 11, p. 73-84.

Ives, R. L., (1935), Fluorine minerals of Colorado: Rocks and Minerals, v. 10, p. 83-84.

Jennings, E. P., (1877), Analyses of tellurium minerals: Am. Inst. Mining Metall. Engineers Trans., v. 6, p. 506-508.

Kelly, W. C., (1958), Mineralogy of limonite in lead-zinc gossans: Econ. Geology, v. 52, p. 536-545.

Kiukkula, K., and Wagner, C., (1957), Measurements on galvanic cells involving solid electrolytes: Jour. Electrochem. Soc., v. 104, p. 385-386.

Klevstov, P. V., and Lemmlein, G. G., (1959), Pressure corrections for the homogenization temperatures of aqueous NaCl solutions: Dokl. Akad. Nauk. S.S.S.R., v. 128, p. 1250-1253.

Kracek, F. C., Ksanda, C. J., and Cabri, L. J., (1966), Phase relations in the system silver-tellurium: Am. Mineralogist, v. 51, p. 14-28.

Kullerud, G., (1962), The Fe-Ni-S system: Carnegie Inst. Washington Yearbook 61, p. 144-150.

Lindgren, W. T., (1907), Some gold and tungsten deposits of Boulder County, Colorado: Econ. Geology, v. 2, p. 453-463.

——, (1933), Mineral deposits: New York, McGraw-Hill, 930 p.

Lindgren, W. T., and Ransome, F. L., (1906), Geology and ore deposits of the Cripple Creek district, Colorado: U.S. Geol. Survey Prof. Paper 54, 516 p.

Locke, A., (1926), Leached outcrops as guides to copper ore: Baltimore (Maryland), Williams and Wilkins, 175 p.

Loughlin, G. F., and Koschman, A. H., (1935), Geology and ore deposits of the Cripple Creek district, Colorado: Colorado Sci. Soc. Proc., v. 13, p. 217-435.

Lovering, T. S., (1941), The origin of tungsten ores of Boulder County, Colorado: Econ. Geology, v. 36, p. 229-279.

——, (1950), The geochemistry of argillic and related types of rock alteration: Colorado School Mines Quart., v. 45, p. 231-260.

Lovering, T. S., and Goddard, E. N., (1938), Laramide igneous sequences and differentiation in the Front Range, Colorado: Geol. Soc. America Bull., v. 49, p. 35-68.

——, (1950), Geology and ore deposits of the Front Range, Colorado: U.S. Geol. Survey Prof. Paper 223, 319 p.

Lovering, T. S., and Tweto, O. L., (1953), Geology and ore deposits of the Boulder County tungsten district, Colorado: U.S. Geol. Survey Prof. Paper 245, 199 p.

Machol, R. E., (1959), Thermodynamic properties of nonstochiometric nickel tellurides and of tellurium: Ph. D. Dissertation, Univ. Michigan, Dissertation Abstracts, v. 20, p. 1193-1194.

Mandarino, J. A., and Mitchell, R. S., (1963a), Spiroffite, a new tellurite mineral from Moctezuma, Sonora, Mexico: Mineralogical Soc. Am. Spec. Paper 1, p. 305-309.

——, (1963b), Denningite, a new tellurite from Moctezuma, Sonora, Mexico: Canadian Mineralogist, v. 7, p. 443-452.

Mandarino, J. A., and Williams, S. J., (1961), Five new minerals from Moctezuma, Sonora, Mexico: Science, v. 133, p. 2017.

Markham, N. L., (1960), Synthetic and natural phases in the system Au-Ag-Te. Parts 1 and 2: Econ. Geology, v. 55, p. 1148-1178, 1460-1475.

Morimoto, N., and Clark, L. A., (1961), Arsenopyrite crystal-chemical relations: Am. Mineralogist, v. 46, p. 1448-1469.

Palache, C., Berman, A., and Frondel, C., (1944), Dana's system of mineralogy. I: New York, John Wiley, 7th ed., 834 p.

Peacock, M. A., and Thompson, R. M., (1946), Montbrayite, a new gold telluride: Am. Mineralogist, v. 31, p. 515-526.

Peter, F., (1923), Uber Brechungsindiges und Absorptionskonstanten der Diamenten zwischen 644 und 226 mμ: Zeitschr. fur Phys., v. 15, p. 358-368.

Posnjak, E., and Merwin, H. E., (1922), The system Fe_2O_3-SO_3-H_2O: Am. Chem. Soc. Jour., v. 44, p. 1965-1994.

Radtke, A. S., (1963), Data on cuprian coloradoite from Kalgoorlie, western Australia: Econ. Geology, v. 58, p. 593-598.

Ramdohr, P., (1950), Die Erzmineralien und ihre Verwachsungen: Berlin, Akademie-Verlag, 826 p.

Ransome, F. L., (1909), The geology and ore deposits of Goldfield, Nevada: U.S. Geol. Survey Prof. Paper 66, 258 p.

———, (1910), The geology and ore deposits of Goldfield, Nevada: Econ. Geology, v. 5, p. 301-311, 438-470.

Roedder, E., (1962), Studies of fluid inclusions I: low temperature application of a dual-purpose freezing and heating stage: Econ. Geology, v. 57, p. 1045-1061.

———, (1963a), Studies of fluid inclusions II: freezing data and their interpretation: Econ. Geology, v. 58, p. 167-211.

———, (1963b), Evidence from fluid inclusions as to the nature of ore-forming fluids; Problems of Post-magmatic Ore Deposition Symposium, Prague, 1965: Geol. Survey Czech. Symposium Vol. 2, p. 375-384.

———, (1965), Report on S. E. G. symposium on chemistry of the ore-forming fluids: Econ. Geology, v. 60, p. 1380-1403.

Rowland, J. F., and Berry, L. G., (1951), The structural lattice of hessite: Am. Mineralogist, v. 36, p. 471-479.

Sato, M., (1960), Oxidation of sulfide ore bodies; I. Oxidation mechanisms of sulfide minerals at 25°C. II. Geochemical environments in terms of Eh and pH: Econ. Geology, v. 55, p. 928-961, 1202-1231.

Schaller, W. T., (1917), On the identity of hamlinite with goyazite: Am. Jour. Sci., v. 43, 4th ser., p. 163-164.

Schmaltz, R. F., (1959), A note on the system Fe_2O_3-H_2O: Jour. Geophys. Research, v. 64, p. 575-579.

Schmitt, H. H., *Editor,* (1962), Equilibrium diagrams for minerals: Cambridge (Massachusetts), Geological Club of Harvard, 199 p.

Schouten, C., (1962), Determination tables for ore microscopy: Amsterdam, Elsevier, 242 p.

Searls, F., Jr., (1948), A contribution to the published information on the geology and ore deposits of Goldfield, Nevada: Univ. Nevada Bull., v. 42, 24 p.

Short, M. N., (1937), Etch tests on calaverite, krennerite, and sylvanite: Am. Mineralogist, v. 22, p. 667-674.

———, (1940), Microscopic determination of the ore minerals: U.S. Geol. Survey, Bull. 914 (2d ed.), 314 p.

Silliman, B., Jr., (1874a), The telluride ores of the Red Cloud and Cold Spring mines, Gold Hill, Colorado: Hayden Survey, 7th Ann. Rept., p. 688-691.

———, (1874b), Mineralogical notes: tellurium ores of Colorado; with a note by A. P. Marvine on the position and geology of the Gold Hill mining region: Am. Jour. Sci., 3d ser., v. 8, p. 25-29.

Sindeeva, N. D., (1964), Mineralogy and types of deposits of selenium and tellurium: New York, John Wiley, 363 p. (Trans. by Geochem. Soc. Am.).

Smith, J. A., (1883), Report on the development of the mineral, metallurgical, agricultural, pastoral, and other resources of Colorado for the years 1881-1882, 159 p. (*see* Eckel, 1961, p. 42, 220.)

Stillwell, F. L., (1931), The occurrence of telluride minerals at Kalgoorlie: Australasian Inst. Mining Metall. Proc. no. 48, p. 115-190.

———, (1949), Occurrence of tellurides at Vatukoula, Fiji: Australasian Inst. Mining Metall. Proc. no. 154-155, new ser., p. 3-27.

Swanson, H. E., and Tatge, E., (1953), Standard X-ray diffraction powder patterns: U. S. Natl. Bur. Standards Circular 539, v. 1, p. 69-70.

Switzer, G., and Swanson, H. E., (1960), Paratellurite, a new mineral from Mexico: Am. Mineralogist, v. 45, p. 1272-1274.

Terziev, G., (1966), Kostovite, a gold-copper telluride from Bulgaria: Am. Mineralogist, v. 51, p. 29-36.

Thompson, J. B., Jr., (1958), Local equilibrium in metasomatic processes: p. 427-457 in Researches in geochemistry, New York, John Wiley, 511 p.

Thompson, R. M., (1947), Frohbergite, FeTe$_2$: a new member of the marcasite group: Univ. Toronto, Geol Ser., v. 50, p. 77-78.

——, (1949), The telluride minerals and their occurrences in Canada: Am. Mineralogist, v. 34, p. 342-382.

Thomson, James E., (1948), Geology of the main ore zone at Kirkland Lake: Ontario Dept. Mines Bull., v. 57, p. 55-103.

Todd, E. W., (1928), Kirkland Lake gold area: Ontario Dept. Mines Bull., v. 37, pt. 2, 176 p.

Tolman, C. F., and Ambrose, J. W., (1934), The rich ores of Goldfield, Nevada: Econ. Geology, v. 29, p. 255-279.

Toulmin, P., III, and Barton, P. B., (1964), A thermodynamic study of pyrite and pyrrhotite: Geochimica et Cosmochimica Acta, v. 28, p. 641-671.

Tunell, G., (1941), The atomic arrangement of sylvanite: Am. Mineralogist, v. 26, p. 457-477.

——, (1954), The crystal structures of the gold-silver tellurides: Office Naval Research Proj. N.R.-081-105, 68 p.

Tunell, G., and Pauling, L., (1952), The atomic arrangement and bonds of the gold-silver ditellurides: Acta Cryst., v. 5, p. 375-381.

Tweto, O. L., (1947), Scheelite in the Boulder district, Colorado: Econ. Geology, v. 42, p. 952-963.

Tweto, O. L., and Sims, P. K., (1963), Precambrian ancestry of the Colorado Mineral Belt: Geol. Soc. American Bull., v. 74, p. 991-1014.

Uytenbogaardt, W., (1951), Tables for microscopic identification of ore minerals: Princeton (New Jersey), Princeton Univ. Press, 242 p.

Van Rensburg, W. C. J., and Cameron, E. N., (1965), Additional data on rotation properties of ore minerals II: Econ. Geology, v. 60, p. 1718-1720.

Van Tuyl, F. M., and Lovering, T. S., (1935), Physiographic development of the Front Range: Geol. Soc. America Bull., v. 46, p. 1291-1350.

Wahlstrom, E. E., (1947), Cenozoic physiographic history of the Front Range, Colorado: Geol. Soc. America Bull., v. 58, p. 551-572.

——, (1950), Melonite in Boulder County, Colorado: Am. Mineralogist, v. 35, p. 948-953.

Warren, H. V., (1947), A new type of gold deposit in British Columbia: Royal Soc. Canada Trans., v. 41, ser. 3, p. 61-72.

Westrum, E. F., Jr., and Machol, R. E., (1958), Thermodynamics of nonstochiometric nickel tellurides and phase relations of tellurium-rich compositions: Jour. Chem. Physics, v. 29, p. 824-828.

Wilkerson, A. S., (1939), Telluride-Tungsten mineralization of the Magnolia mining district, Colorado: Econ. Geology, v. 34, p. 437-450.

Williams, D., and Nakhla, F. M., (1951), Chromagraphic contact print method of examining metallic minerals and its applications: Inst. Mining and Metallurgy Trans., v. 60, p. 257-295.

Plates

TYPICAL ORE SHOWING TELLURIDE-BEARING HORN QUARTZ SEAMS

Polished specimen of typical ore showing network of telluride-bearing horn quartz seams. Reflected light, in air, ¾ X. Interocean vein.

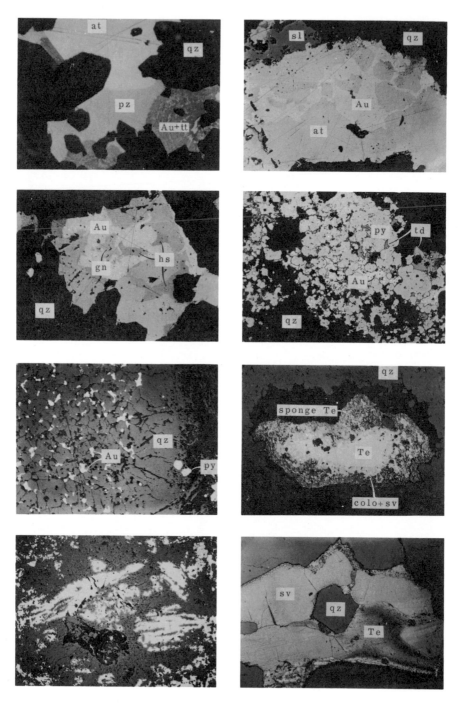

PLATE 2.

a. Altaite (at) and petzite (pz) in quartz (qz). Concentric aggregate of fine-grained gold in tellurium oxide (Au + tt) selectively replaced petzite. Reflected light, in air, 250 X. Horsefal vein.

b. Altaite (at) veined by native gold (Au). Relict of sphalerite (sl) appears at edge of vug. Reflected light, in air, 183 X. Smuggler mine.

c. Galena (gn) replaced by hessite (hs) which was in turn veined by native gold (Au). Reflected light, in air, 218 X. Nancy mine.

d. Tetrahedrite (td) interstitial to pyrite (py) has been replaced by native gold (Au). Reflected light, in air, 184 X. Colorado vein.

e. Small grains of gold (Au) dispersed through cavities in sugary vein quartz (qz). Reflected light, in air, 153 X. Croesus mine.

f. Single crystal of tellurium (Te) was largely replaced by intergrown coloradoite (colo) and sylvanite (sv) and outer layer of sponge tellurium (sponge Te). Outlines of original crystal visible. Reflected light, in air, 157 X. Emancipation mine.

g. Subhedral, bladed tellurium in penecontemporaneous fine-grained quartz. Reflected light, in air, 28 X. John Jay mine.

h. Sylvanite (sv) and native tellurium (Te) in quartz (qz). In places, tellurium veins the sylvanite. (Section immersed about 2 seconds in conc. HNO_3 producing etch cleavage in sylvanite and dark differential coloration of tellurium.) Reflected light, in air, 176 X. Alpine Horn mine.

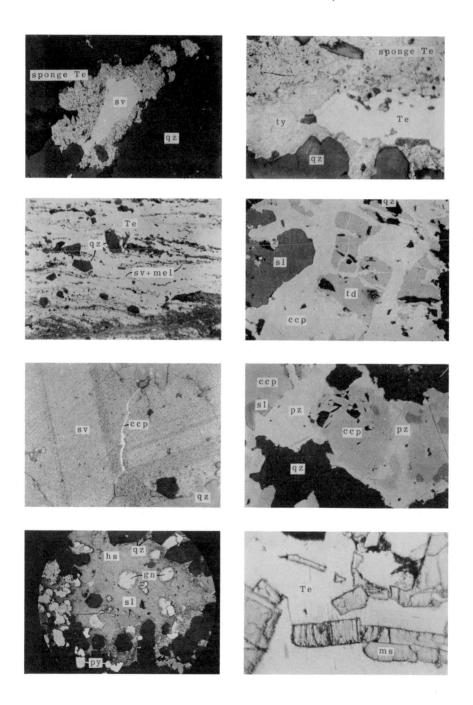

PLATE 3.

a. Single twinned crystal of sylvanite (sv) in partially open vug has been marginally replaced by sponge tellurium (sponge Te). The matrix is quartz (qz). Reflected light, in air, 218 X. Bumble Bee mine.

b. Single tellurium (Te) crystal replaced by tetradymite (ty) which was in turn replaced by sponge tellurium (sponge Te). Matrix is quartz (qz). (Field etched 20 seconds with 5 percent HCl which selectively attacked tetradymite.) Reflected light, in air, 265 X. Emancipation mine.

c. Brecciated quartz (qz) and streaks of sylvanite (sv), melonite (mel), and gangue strung out in deformed native tellurium (Te). Reflected light, in air, 158 X. White Crow mine.

d. Sphalerite (sl) and tetrahedrite (td) veined and replaced by chalcopyrite (ccp). Gangue is quartz (qz). Reflected light, in air, 212 X. Monitor tunnel.

e. Veinlet of chalcopyrite (ccp) along grain boundaries in sylvanite (sv). (Section etched about 20 seconds with 1:7 HNO_3.) Reflected light, in air, 394 X. Bondholder mine.

f. Sphalerite (sl) and chalcopyrite (ccp) veined and replaced by petzite (pz). Gangue is quartz (qz). Reflected light, in air, 170 X. Monitor tunnel.

g. Round relicts of sphalerite (sl) and galena (gn) in mottled hessite. Numerous inclusions of tetrahedrite produce the mottling in hessite. Grains of pyrite (py) also appear in hessite (hs) and surrounding quartz (qz). Reflected light, in air, 89 X. Colorado vein.

h. Slightly rounded blades of marcasite (ms) enclosed and in places veined by native tellurium (Te). Reflected light, in oil, 386 X. Slide vein.

PLATE 4.

a. Early bent crystals of marcasite (ms) encased in quartz (qz). Altaite (at) with inclusions of Coloradoite (colo) fills vug in the quartz. Reflected light, in air, 126 X. John Jay mine.

b. Colloform marcasite in vein quartz which carries scattered grains of pyrite. Reflected light, in air, 11 X. Colorado vein.

c. Plumose marcasite (ms) coated by fine-grained pyrite (py) and intergrown with quartz (qz) at edge of vug. The vug is filled by petzite (pz). Reflected light, in air, 153 X. Nancy mine.

d. Vug in quartz (qz) is lined by early pyrite (py), bravoite (bv), and melonite (mel), and filled by late hessite (hs) which locally veins the pyrite. Reflected light, in oil, 415 X. Osceola-Interocean mine.

e. Deformed flakes of molybdenite and sparse grains of pyrite (py) in early vein quartz. Reflected light, in air, 154 X. Poorman mine.

f. Stromeyerite (strom) in calcite (calc) that coats older quartz (qz). See also Plate 13c. Early pyrite (py) was in places corroded by hessite (hs) and gold (Au) which are concentrated in spaces between quartz crystals. Reflected light, in air, 30 X. Colorado vein.

g. Scalloped inclusions of coloradoite (colo) and sylvanite (sv) in altaite (at). Reflected light, in air, 99 X. John Jay mine.

h. Narrow rim of sylvanite (sv) between calaverite (ca) and altaite (at). Supergene rickardite (rk) appears at contacts of sylvanite and calaverite with quartz. Reflected light, in air, 675 X. Horsefal vein.

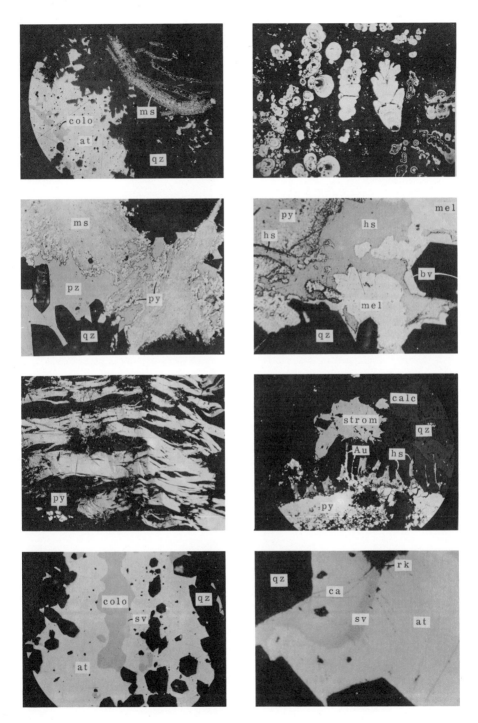

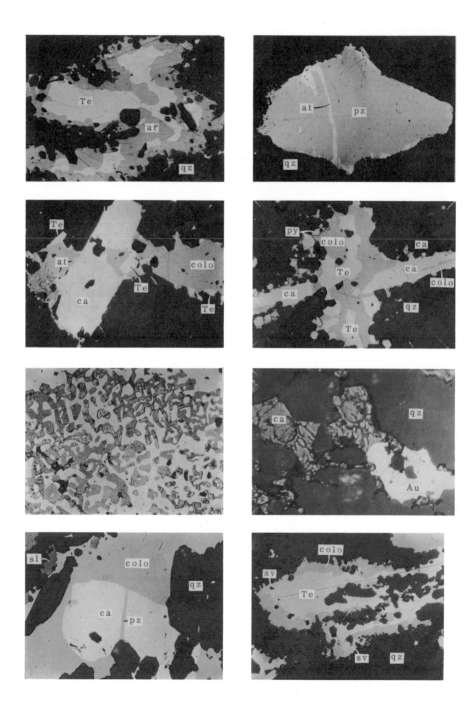

PLATE 5.

a. Late altaite (at) along the borders of coarsely crystalline tellurium (Te) in quartz (qz). (Altaite etched by fumes of conc. HNO_3.) Reflected light, in air, 112 X. John Jay mine.

b. Altaite (at) veins vug filling of petzite (pz) in quartz (qz). Reflected light, in air, 182 X. Slide vein.

c. Early calaverite crystal (ca) extends across vug in quartz. Younger coloradoite (colo), tellurium (Te), and altaite (at) occur within the vug. Reflected light, in air, 189 X. John Jay mine.

d. Subhedral blade of calaverite (ca) in contact with coloradoite (colo) and tellurium (Te) which are anhedral and largely confined to vugs in quartz (qz). Reflected light, in air, 142 X. John Jay mine.

e. Subgraphic inclusions of calaverite in coloradoite. Groups of adjacent inclusions are optically oriented parts of calaverite crystals. (Calaverite was etched 30 seconds with 1:7 HNO_3.) Reflected light, in air, 186 X. Emancipation mine.

f. Calaverite (ca) and native gold (Au) in quartz (qz). (Field etched 1 minute with 1:7 HNO_3; circular areas in calaverite are relatively unetched and mark locations of standing bubbles during effervescence.) Reflected light, in air, 492 X. Nancy mine.

g. Calaverite (ca) veined by petzite (pz) and both minerals surrounded by coloradoite (colo). Relict grains of sphalerite (sl) appear on walls of vug in quartz (qz). Reflected light, in air, 238 X. Logan mine.

h. Rims of coloradoite (colo) and minor sylvanite (sv) border single anhedral crystals of tellurium (Te). All tellurium in center of photograph extinguishes simultaneously. Reflected light, in air, 218 X. John Jay mine.

PLATE 6.

a. Partial pseudomorph of petzite (pz) and coloradoite (colo) after bladed sylvanite. Relicts of sylvanite (sv) are optically aligned. (Petzite etched by $HgCl_2$.) Reflected light, in oil, 340 X. Logan mine.

b. Coloradoite (colo) borders and has evidently replaced petzite (pz). An unidentified silver sulfantimonide (B) occurs in petzite and apparently guided replacement of coloradoite. (Section etched 1 minute with 5 percent HNO_3.) Reflected light, in air, 215 X. King Wilhelm mine.

c. Coloradoite (colo) rims and crosscuts older sylvanite (sv) and both minerals are surrounded by older pyrite (py). The pyrite is veined by supergene goethite (gt) and patches of rusty gold (Au + gt) occur along coloradoite-sylvanite contacts. Reflected light, in air, 139 X. Poorman mine.

d. Mutual intergrowth of coloradoite (colo) and empressite (epr). Strong bireflectance of empressite apparent in photograph. Black areas are subgraphic voids in empressite. Reflected light, in oil, 169 X. Empress mine.

e. Gridwork of inversion (?) twin lamellae typical of hessite (hs). The untwinned mineral is petzite (pz). Reflected light, in oil, 375 X. Analyzer in and set at 88°. Cold Spring-Red Cloud mine.

f. Patches of untwinned hessite (hs_2) appear within twinned hessite (hs_1). The untwinned grains are crystallographically aligned. Reflected light, in oil, 510 X. Analyzer in and set at 88°. Cold Spring-Red Cloud mine.

g. Sylvanite (sv) is veined and replaced by hessite (hs). Lamellar growth twins are visible in bireflectant sylvanite. Reflected light, in air, 176 X. Osceola-Interocean mine.

h. Early marcasite (ms) is coated by pyrite (py) and intergrown with quartz (qz) along the walls of vug. Petzite (pz) in mutual contact with hessite (hs)-sylvanite (sv) intergrowth comprises late vug filling. Reflected light, in oil, 169 X. Nancy mine.

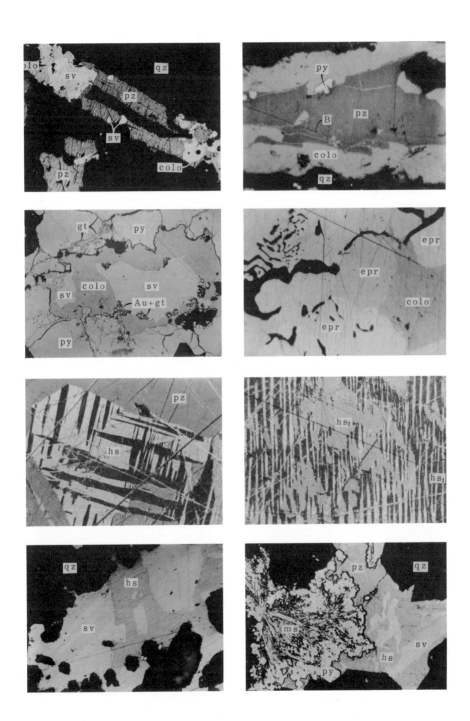

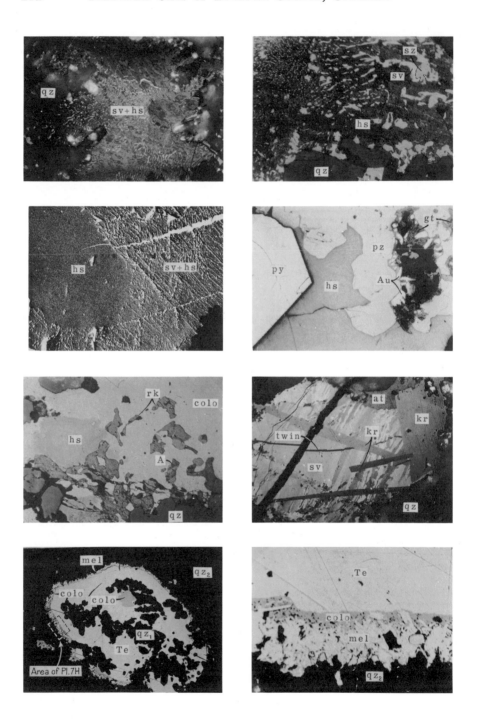

PLATE 7.

a. Micrographic intergrowth of sylvanite in hessite (sv + hs). The hessite shows irregular inversion (?) twinning. As stage is rotated, numerous other inclusions of sylvanite are illuminated. Reflected light, in air, 199 X. Crossed polars. Cash mine.

b. Composite stuetzite (sz)-sylvanite (sv) inclusions in hessite (hs). Plate 7a represents another part of this same polished surface. (The hessite is etched black by brief immersion of section in 1:7 HNO_3.) Grains of quartz (qz) appear at base of photograph. Reflected light, in air, 270 X. Cash mine.

c. Electron photomicrograph of replica of hessite-sylvanite intergrowth (sv + hs) shown in Plate 7a. Sylvanite appears white. It veins an inclusion-free phase that is probably also hessite (hs). The inclusions of sylvanite in hessite persist below the limits of optical resolution. 8200 X. Cash mine.

d. Mutual intergrowth of hessite (hs) and petzite (pz) encloses subhedral pyrite (py). Supergene cavities are lined by goethite (gt) and contain wires of native gold (Au). (The hessite was etched dark gray by brief exposure to 5 percent HNO_3.) Reflected light, in air, 50 X. Cold Spring-Red Cloud mine.

e. Inclusions of rickardite (rk) in unoxidized coloradoite (colo). Flecks and veinlets of native gold (Au) are barely visible in the rickardite. The cloudy gray area (hs) is fine-grained supergene hessite that replaced coloradoite. Relict lamellar twinning in some grains of rickardite (grain labelled A) suggests original sylvanite. Reflected light, in air, 156 X. Last Chance mine.

f. Deformed sylvanite (sv) replaced by thin blades of krennerite (kr) and both minerals in turn replaced by late altaite (at). Thin primary twin at extinction in sylvanite shows offsets particularly along secondary twin lamellae. Fine-grained recrystallized sylvanite also present. Crack produced in mounting (black) cuts across the specimen. Reflected light, in air, 194 X. Alpine Horn mine.

g. Replacement rims of melonite (mel) and coloradoite (colo) along the borders of a single tellurium crystal (Te). The crystal is mounted on early vein quartz (qz_1) and surrounded by younger vein quartz (qz_2). Reflected light, in air, 36 X. John Jay mine.

h. Enlargement of small area indicated in Plate 7g showing details of melonite-coloradoite rims. The quartz visible (qz_2) is the later generation younger than tellurium (see Pl. 7g). The melonite (mel) encloses angular bits of this quartz and apparently formed after the tellurium was surrounded by the quartz. Reflected light, in oil, 410 X. John Jay mine.

PLATE 8.

a. Aggregate of fine-grained melonite (mel) pseudomorphic after bladed tellurium in quartz (qz). No tellurium is preserved in the field of view. Reflected light, in air, 61 X. Eclipse mine.

b. Common zonal arrangement of minerals along quartz (qz)-tellurium (Te) contacts. Melonite (mel) on quartz is separated from tellurium by narrow band of intergrown sylvanite (sv) and coloradoite (colo). Altaite (at) (etched black by dilute HCl) occurs along melonite-coloradoite contact. Reflected light, in oil, 278 X. Lady Franklin mine.

c. Melonite (mel) appears in late petzite (pz) that veined and replaced calaverite (ca). The melonite is localized on quartz (qz) and roscoelite (rosc) surfaces. Reflected light, in air, 198 X. King Wilhelm mine.

d. Petzite (pz) fills vug in quartz (qz) lined by early pyrite (py) and melonite (mel). Reflected light, in air, 69 X. Osceola-Interocean mine.

e. Melonite crystal (mel) in vug in quartz (qz) is cemented by hessite. Melonite in vug to the right contains round inclusions of chalcopyrite (ccp). Reflected light, in air, 96 X. Winona mine.

f. Large single crystal of melonite (mel) attached to quartz (qz) wall of vug and surrounded by younger hessite (hs). Reflected light, in air, 160 X. Winona mine.

g. Coarse fibers of melonite (mel) lining vug in quartz (qz) that is filled by coloradoite (colo). Thin film of quartz (qz) formed on melonite before introduction of coloradoite. A fine layer of altaite (at) borders the coloradoite. Reflected light, in air, 320 X. John Jay mine.

h. Nagyagite (ng) intergrown with petzite (pz) in vug in quartz (qz) lined by pyrite (py). Petzite follows cleavages in the nagyagite. (Section etched briefly with both dilute HNO$_3$ and dilute HCl.) Reflected light, in air, 238 X. Nancy mine.

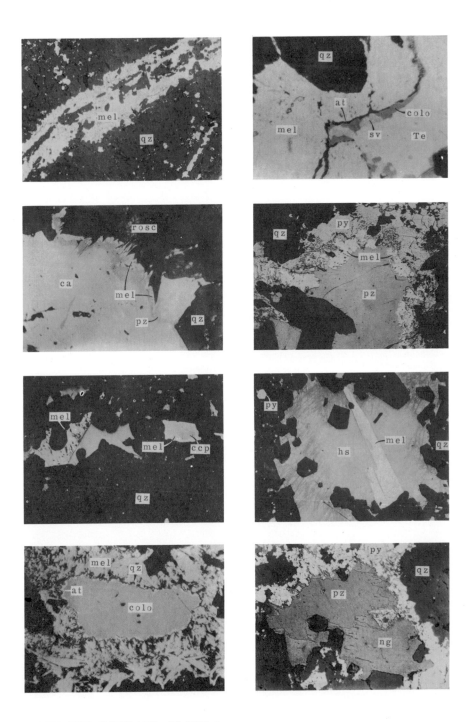

KELLY AND GODDARD, PLATE 8
Geological Society of America Memoir 109

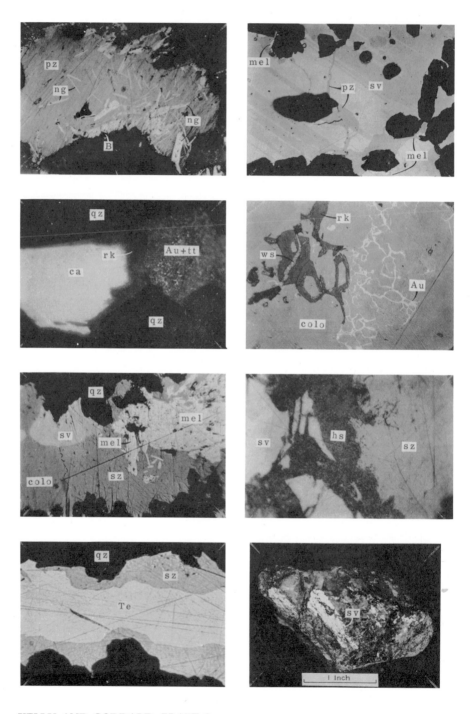

KELLY AND GODDARD, PLATE 9
Geological Society of America Memoir 109

PLATE 9.

a. Numerous crystals of nagyagite (ng) scattered through petzite (pz). An unidentified silver sulfantimonide (B) coats walls of vug and was in part replaced by the petzite. (Field etched about 15 seconds with 1:7 HNO₃.) Reflected light, in air, 214 X. King Wilhelm mine.

b. Twinned sylvanite (sv) is veined by petzite (pz). Sparse melonite (mel) occurs in petzite at quartz (qz) contacts. Reflected light, in air, 178 X. Ingram mine.

c. Calaverite (ca) was marginally replaced by rickardite (rk). Intergrowth of tellurite (tt) and fine-grained gold (Au) probably marks site of original petzite. Gangue is quartz (qz). Reflected light, in oil, 737 X. Horsefal vein.

d. Inclusions of intergrown rickardite (rk) and native gold (Au) in coloradoite. Supergene gold also veins coloradoite. Veinlet of intergrown gold and weissite (?) (ws) cuts one rickardite inclusion. Reflected light, in oil, 494 X. Last Chance mine.

e. Complex vug filling of melonite (mel), stuetzite (sz), coloradoite (colo), and sylvanite (sv) in quartz (qz). Reflected light, in oil, 330 X. Empress mine.

f. Stuetzite (sz) and sylvanite (sv) are veined by hessite (hs). Hessite etched by 1:7 HNO₃. Reflected light, in oil, 840 X. Black Rose mine.

g. Single crystal of tellurium (Te) bordered by stuetzite (sz). Gangue is quartz (qz). Reflected light, in oil, 284 X. Smuggler mine.

h. Hand specimen split along seam of telluride-bearing quartz. Thin blades of sylvanite (sv) are "plastered" on the fine-grained quartz. Ingram mine.

PLATE 10.

a. Blades of sylvanite in fine-grained quartz. Reflected light, in air, 11 X. Ingram mine.

b. Petzite (pz) veins sylvanite (sv) and adjacent tetrahedrite (td). The sylvanite and tetrahedrite formed as interlocking crystals in open vugs. Black areas in the photograph are voids. Reflected light, in oil, 832 X. Poorman mine.

c. Bladed pseudomorphs of intergrown altaite and petzite (indistinguishable in photograph) after original sylvanite. The main mineral in the field is altaite. See also Figure 10. Reflected light, in air, 9 X. Smuggler mine.

d. Small grains of sylvanite (sv) along grain boundaries in petzite (pz). Reflected light, in air, 198 X. Smuggler mine.

e. Sylvanite crystals (sv) intergrown with coloradoite (colo) along the quartz walls (qz) of a vug filled by native tellurium (Te). (Section etched several seconds in conc. HNO$_3$ producing etch cleavage in sylvanite and dark coloration of tellurium.) Reflected light, in air, 178 X. Emancipation mine.

f. Tellurium (Te) apparently replaced by coloradoite (colo) with numerous grains of sylvanite (sv) concentrated in and along coloradoite "front." Reflected light, in air, 139 X. Lady Franklin mine.

g. Veinlet of petzite (pz) and fine-grained sylvanite (sv) along offset in coarsely crystalline sylvanite (sv). Most sylvanite is at extinction but several offsets of illuminated primary twins are visible. Reflected light, in air, 41 X. Crossed polars. Alpine Horn mine.

h. Veinlet of petzite (pz) and fine-grained, twinned sylvanite (sv) transecting coarsely crystalline calaverite (ca). Reflected light, in oil, 820 X. Buena mine.

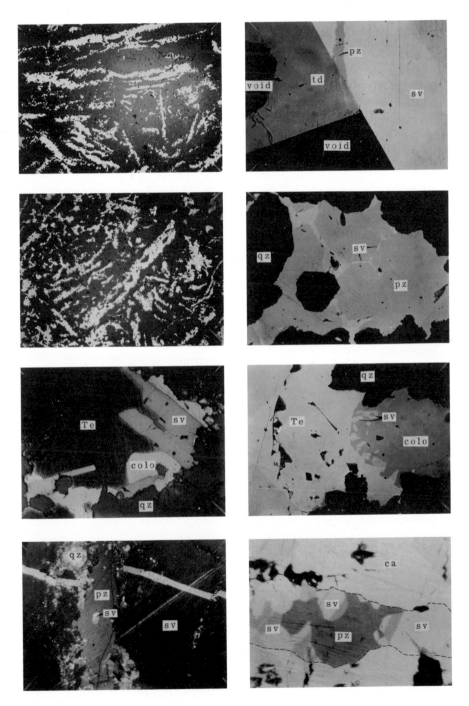

KELLY AND GODDARD, PLATE 10
Geological Society of America Memoir 109

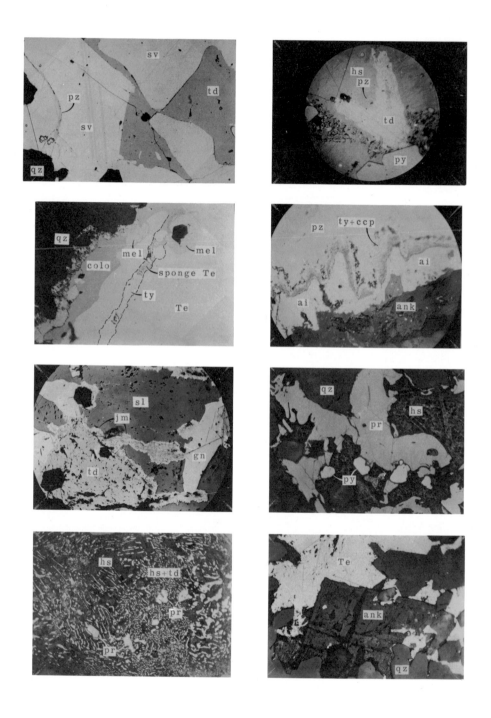

PLATE 11.

a. Intergrowth of contemporaneous sylvanite (sv) and tetrahedrite (td). Petzite (pz) veins the sylvanite and elsewhere in this sample (see Pl. 10b) veins both sylvanite and tetrahedrite. Reflected light, in air, 204 X. Poorman mine.

b. Crystals of tetradymite (ty) extend across hessite (hs)-petzite (pz) intergrowth. The pitted area in photograph is a fine-grained intergrowth of aikinite, tetradymite, and chalcopyrite. Pyrite (py) crystals appear at base of photograph. Reflected light, in air, 30 X. Analyzer in and set at 85°. Cold Spring-Red Cloud mine.

c. Veinlet of tetradymite (ty) in tellurium (Te) connects with areas of coloradoite (colo) and melonite (mel) along quartz (qz) surfaces. The pitted areas within the tetradymite along quartz contacts are sponge tellurium. Reflected light, in oil, 317 X. Lady Franklin mine.

d. Petzite (pz) replaced by aikinite (ai) with zone of intergrown tetradymite and chalcopyrite (ty + ccp) along their contacts. The aikinite is attached to late ankerite (ank) seen at the base of the photograph. Reflected light, in oil, 504 X. Cold Spring-Red Cloud mine.

e. Thin, fibrous layers of jamesonite (jm) are localized along contacts of tetrahedrite (td) with sphalerite (sl). Contacts of late galena (gn) lack jamesonite. Reflected light, in air, 89 X. Poorman mine.

f. Irregular inclusions of pyrargyrite (pr) in hessite (hs) in vug in quartz (qz). (Twinning visible in hessite which was etched by immersion for about 20 seconds in 5 percent HNO_3.) Reflected light, in air, 257 X. Gray Copper mine.

g. Eutectoid intergrowth of pyrargyrite (pr) and tetrahedrite (td) in hessite (hs). (Hessite selectively etched by immersion for about 30 seconds in 1:7 HNO_3.) Reflected light, in air, 197 X. Colorado vein.

h. Early crystal of ankerite (ank) on walls of vug in quartz (qz) which is filled by native tellurium (Te). Reflected light, in air, 158 X. Richmond mine.

PLATE 12.

a. Crystals of ferberite (fb) in quartz (qz). Petzite (pz) and native gold (Au) occupy vugs and locally corroded the ferberite. Reflected light, in air, 128 X. Kekionga mine.

b. Euhedral quartz crystal (qz) has been corroded by later sulfides, sphalerite (sl), and chalcopyrite (ccp). Hessite (hs) occurs along contacts of older quartz and sulfides. Reflected light, in air, 168 X. American mine.

c. Curious network of fractures in fine-grained vein quartz. Note that tellurides (chiefly altaite, petzite, sylvanite) fill some fractures and by-pass others. Reflected light, in air, 10 X. Smuggler mine.

d. Calaverite (ca) apparently was replaced by petzite (pz) and coloradoite (colo) in vug in quartz (qz). Fine-grained gold (Au) occurs in goethite (gt) as an oxidation product of the tellurides. Reflected light, in air, 230 X. Poorman mine.

e. Supergene tellurite (tt) replaced native tellurium (Te) in quartz (qz). Intergrown rickardite and gold (rk + Au) probably mark the location of original gold telluride grains. Reflected light, in air, 99 X. Potato Patch mine.

f. Polished thin section of quartz from telluride ore showing two-phase (vapor + liquid) H_2O type fluid inclusion. Transmitted light, in air, 1500 X. Gray Eagle mine.

g. Polished thin section of fluorite from pyritic gold ore showing two-phase (vapor + liquid) H_2O type inclusion. "Meniscus" in vapor phase (as in Pl. 12f) is produced by contact of bubble with walls of inclusion and not by a third phase. Transmitted light, in air, 1710 X. Stanley mine.

h. Polished thin section of quartz from pyritic gold ore showing two-phase (vapor + liquid) H_2O type inclusion. Transmitted light, in air, 1720 X. Stanley mine.

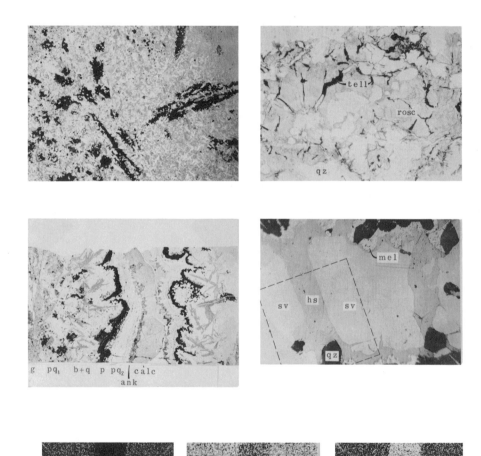

PLATE 13.

a. Thin section showing relationship of fine-grained vein quartz to included elongate crystals of native tellurium. (Analyzer is in but not quite crossed.) Transmitted light, in air, 14 X. John Jay mine.

b. Thin section showing tellurides (tell) that selectively replaced roscoelite (rosc) in quartz (qz)-roscoelite intergrowth. Transmitted light, in air, 11 X. King Wilhelm mine.

c. Thin section showing mineral zoning in banded ore. The labeled zones are altered granite (g), early pyritic-quartz (pq_1), barite plus quartz (b+q), pyrite (p), late pyritic quartz (pq_2), ankerite (ank), and calcite (calc). Transmitted light, in air, 3.3 X. Colorado vein.

d. Veinlets of hessite (hs) in sylvanite (sv) contain fine-grained melonite (mel) that is concentrated along quartz (qz) surfaces. The outlined area is that analyzed in Plates 13e, 13f, and 13g. Reflected light, in air, 123 X. Ingram mine.

e. Electron microprobe image showing the distribution of gold in the area outlined in Plate 13d. 170 X. Ingram mine.

f. Electron microprobe image showing the distribution of tellurium in the area outlined in Plate 13d. 170 X. Ingram mine.

g. Electron microprobe image showing the distribution of silver in the area outlined in Plate 13d. 170 X. Ingram mine.

Index

227